The 1984
Olympic Scientific
Congress
Proceedings
Volume 1

Perspectives in Kinanthropometry

Series Editors:

Jan Broekhoff, PhD
Michael J. Ellis, PhD
Dan G. Tripps, PhD

University of Oregon
Eugene, Oregon

The 1984
Olympic Scientific
Congress
Proceedings
Volume 1

Perspectives in Kinanthro- pometry

James A. P. Day
Editor

Human Kinetics Publishers, Inc.
Champaign, Illinois

Library of Congress Cataloging-in-Publication Data

Olympic Scientific Congress (1984 : Eugene, Or.)
   Perspectives in kinanthropometry.

   (1984 Olympic Scientific Congress proceedings ; v. 1)
   Bibliography: p.
   1. Kinesiology—Congresses.   2. Anthropometry—
Congresses.   3. Sports—Physiological aspects—Congresses.
4. Somatotypes—Congresses.   I. Day, James A. P.,
1931-   .   II. Title.   III. Title: Kinanthropometry.
IV. Series: Olympic Scientific Congress (1984 : Eugene,
Or.).   1984 Olympic Scientific Congress proceedings ;
v. 1.
GV565.O46   1984 vol. 1   796 s   85-18118
[QP303]   [612'.044]
ISBN 0-87322-008-0

Managing Editor: Susan Wilmoth, PhD
Developmental Editor: Gwen Steigelman, PhD
Production Director: Sara Chilton
Copyeditor: Carol Poto
Typesetter: Sandra Meier
Text Layout: Cyndy Barnes
Cover Design and Layout: Jack Davis
Printed By: Braun-Brumfield, Inc.

ISBN: 0-87322-006-4   (10 Volume Set)
ISBN: 0-87322-008-0

Printed in the United States of America

10 9 8 7 6 5 4 3 2 1

Human Kinetics Publishers, Inc.
Box 5076, Champaign, IL 61820

Contents

Series Acknowledgments

The Congress organizers realize that an event as large and complex as the 1984 Olympic Scientific Congress could not have come to fruition without the help of literally hundreds of organizations and individuals. Under the patronage of UNESCO, the Congress united in sponsorship and cooperation no fewer than 64 national and international associations and organizations. Some 50 representatives of associations helped with the organization of the scientific and associative programs by coordinating individual sessions. The cities of Eugene and Springfield yielded more than 400 volunteers who donated their time to make certain that the multitude of Congress functions would progress without major mishaps. To all these organizations and individuals, the organizers express their gratitude.

A special word of thanks must also be directed to the major sponsors of the Congress: the International Council of Sport Science and Physical Education (ICSSPE), the United States Olympic Committee (USOC), the International Council on Health, Physical Education and Recreation (ICHPER), and the American Alliance for Health, Physical Education, Recreation and Dance (AAHPERD). Last but not least, the organizers wish to acknowledge the invaluable assistance of the International Olympic Committee (IOC) and its president, Honorable Juan Antonio Samaranch. President Samaranch made Congress history by his official opening address in Eugene on July 19, 1984. The IOC further helped the Congress with a generous donation toward the publication of the Congress papers. Without this donation it would have been impossible to make the proceedings available in this form.

Finally, the series editors wish to express their thanks to the volume editors who selected and edited the papers from each program of the Congress. Special thanks to James A.P. Day of the University of Lethbridge for his work on this volume.

Jan Broekhoff,
Michael J. Ellis, and
Dan G. Tripps

Series Editors

Series Preface

Perspectives in Kinanthropometry contains selected proceedings from this disciplinary program of the 1984 Olympic Scientific Congress, which was held at the University of Oregon in Eugene, Oregon, preceding the Olympic Games in Los Angeles. The Congress was organized by the College of Human Development and Performance of the University of Oregon in collaboration with the cities of Eugene and Springfield. This was the first time in the history of the Congress that the event was organized by a group of private individuals, unaided by a federal government. The fact that the Congress was attended by more than 2,200 participants from more than 100 different nations is but one indication of its success.

The Congress program focused on the theme of Sport, Health, and Well-Being and was organized in three parts. The mornings of the eight-day event were devoted to disciplinary sessions, which brought together specialists in various subdisciplines of sport science such as sport medicine, biomechanics, sport psychology, sport sociology, and sport philosophy. For the first time in the Congress' history, these disciplinary sessions were sponsored by the national and international organizations representing the various subdisciplines. In the afternoons, the emphasis shifted toward interdisciplinary themes in which scholars and researchers from the subdisciplines attempted to contribute to crossdisciplinary understanding. In addition, three evenings were devoted to keynote addresses and presentations, broadly related to the theme of Sport, Health, and Well-Being.

In addition to the scientific programs, the Congress also featured a number of associative programs with topics determined by their sponsoring organizations. Well over 1,200 papers were presented in the various sessions of the Congress at large. It stands to reason, therefore, that publishing the proceedings of the event presented a major problem to the organizers. It was decided to limit proceedings initially to interdisciplinary sessions which drew substantial interest from Congress participants and attracted a critical number of high-quality presentations. Human Kinetics Publishers, Inc. of Champaign, Illinois, was selected to produce these proceedings. After considerable deliberation, the following interdisciplinary themes were selected for publication: Competitive Sport for Children and Youths; Human Genetics and Sport; Sport and Aging; Sport and Disabled Individuals; Sport and Elite Performers; Sport, Health, and Nutrition; and Sport and Politics. The 10-volume set published by Human Kinetics Publishers is rounded out by the disciplinary proceedings of Kinanthropometry, Sport Pedagogy, and the associative program on the Scientific Aspects of Dance.

Jan Broekhoff,
Michael J. Ellis, and
Dan G. Tripps

Series Editors

Preface

At the 1984 Olympic Scientific Congress (OSC), kinanthropometry was recognized as a major scientific discipline. At a time of apparent worldwide acceptance of the name kinanthropometry, it is amusing to note that the Eugene newspaper suggested in a featured article, "You don't have to know how to pronounce, spell, or define kinanthropometry to attend the Olympic Scientific Congress." Perhaps the term still has some mystical connotation for the larger public, but for academicians it should be clear that *kinanthropometry* has been defined and accepted as "the study of human size, shape, proportion, composition, maturation, and gross function, in order to understand growth, exercise, performance, and nutrition." Similar to the mechanistic approach to human motion, anthropometry has a rich tradition within sport science and physical education. For example, early investigators of the physique of Olympic athletes were essentially *kinanthropometrists*, although the word *kinanthropometry* was not part of their daily vocabulary. They tried to relate body structure to the specialized functions needed for various tasks and to understand the limitations of such relationships. The field of kinanthropometry, however, goes beyond the measurement of structural characteristics of the human being to include aspects such as maturation, nutrition, and body composition.

For younger colleagues, but also for historical reasons, it is perhaps interesting to note that the term *kinanthropometry* in its present connotation was first used by Bill Ross in 1972 in the Belgian journal *Kinanthropologie*. Its use spread during the middle and late 1970s. Kinanthropometry was first included in the OSC at Quebec in 1976, prior to the Montreal Olympic Games. Two years later, kinanthropometry had aroused enough interest in the scientific community to warrant a successful Second International Congress on Kinanthropometry, held in Leuven, Belgium. In its short history, kinanthropometry has often been a featured section of major national and international congresses. A very important point in the development of kinanthropometry was the founding (Brasilia, 1978) of the International Working Group on Kinanthropometry (IWGK), under the auspices of the Research Committee of the International Council of Sport Science and Physical Education (ICSSPE; NGO Status A at UNESCO). This group has been the driving force behind numerous initiatives aimed at, among other things, improving the rigor and quality of scientific research in this discipline and enabling persons from all over the world to share their knowledge.

The 27 papers included in this volume were selected from more than 60 presented in the Kinanthropometry section of the 1984 OSC. Their authors represent all six continents and fairly indicate the dispersion of kinanthropometric research around the world.

The papers have been grouped into five "perspectives" loosely focused around the theme presentations by invited speakers William Stini, Robert Malina, Han Kemper, Eduardo DeRose, and William Ross. A perspective can be a device to aid one's vision, a viewpoint, a mental view, or the capacity to view true relationships or judge relative importance. The title of this volume was selected because we like the idea of science (any science) providing a perspective which allows us a better view of things. The areas of inquiry labeled physical anthropometry, maturation, growth and development, and motor development all have advanced because the perspective of kinanthropometry has sharpened the focus in these disciplines. In recent decades (perhaps since Sheldon), the techniques we have labeled kinanthropometry have helped sharpen the focus on the physical performance of athletes and dancers. A large fraction of today's work in kinanthropometry deals with physical performance, especially the maximal performance embodied in sport.

Many people contributed to the preparation of these proceedings. The IWGK must be recognized, especially its chairman, Jan Borms. Jan Broekhoff, the Scientific Program Chairman of the OSC, has been a major force. The demanding task of reviewing the papers was shared by Jan Borms, Lindsay Carter, Dayna Daniels, Don Drinkwater, William Duquet, Marcel Hebbelinck, Lawrence Hoye, Michael Marfell-Jones, Alan Martin, and Richard Ward. At The University of Lethbridge, a corps of typists, including Eileen Ferguson, Susanne Menard, Charlene Sawatsky, and Rita Zaugg, contributed their talents. At Human Kinetics Publishers, Sue Wilmoth and Gwen Steigelman have shown marvelous cooperation and absolutely awesome patience.

James Day
Editor

Jan Borms, Chairman
International Working Group on Kinanthropometry

PART I

*An Anthropological
Perspective*

Part I includes just one paper, William Stini's treatise on kinanthropometry from an anthropological viewpoint. He points out that the condition of the human species is the result of adaptation, both genotypic and phenotypic. Phenotypic adaptations, or "developmentally-acquired traits," are comparable to the changes produced by athletic training, according to Stini. It is this point at which the interests of the physical anthropologist and the kinanthropometrist come together.

We are aware that the skills of anthropometry have been practiced by physical anthropologists for many decades. Jan Borms, in his preface to this volume, reminds us that the term *kinanthropometry* is a young one. It should be clear, however, that what is young is the terminology and not the use of anthropometry to study the relationship of form and function, the "interactions of anatomy, growth, and performance" in Stini's words. If it is true that anthropologists have always been interested in the contributions of form to function, then they, not the sport scientists, are the earlier kinanthropometrists. It is thus entirely appropriate that anthropologist William Stini should have been invited to make a keynote presentation to the Kinanthropometric section of the 1984 Olympic Scientific Congress.

1

Kinanthropometry: An Anthropological Focus

William A. Stini
UNIVERSITY OF ARIZONA
TUCSON, ARIZONA, USA

Humans have probably always been fascinated by the phenomenon of human variation. The ascription of constitutional bases for psychological characteristics can be found in the literature of many nations at many times in history. Such terms as *phlegmatic* and *sanguine* are still used to describe the disposition of an individual, although those who use them are frequently unaware of their original morphotypic referrents. Almost 2,500 years ago, Hippocrates referred to humoral types in his effort to explain relationships between body build and behavior. In the last half of the 19th century, anatomists such as Beneke sought to describe constellations of traits associated with pathological conditions involving several physiological systems simultaneously (Beneke, 1878, 1881). In the early part of the present century, considerable attention was paid to the occurrence of constitutional types (Kretschmer, 1921) and their behavioral correlates (Viola, 1935, 1936).

The development of somatotype analysis by Sheldon and his colleagues (Sheldon, Dupertuis, & McDermott, 1954; Sheldon, Hartl, & McDermott, 1949; Sheldon & Stevens, 1942; Sheldon, Stevens, & Tucker, 1940; Sheldon & Tucker, 1940) stimulated a great deal of interest in the varieties of human physique and methods of quantification. Tanner (1956) applied techniques of somatotyping in his search for predisposing factors in certain diseases, and Brozek (1965) explored the relationship of somatotypes to body composition in great depth. Development of refinements in the somatotype system by Heath and Carter (1967) have maintained interest in the method. In a number of publications over the years, Tanner has sought to identify functional relationships between the rate and time of maturation (Tanner, 1963; Tanner & Inhelder, 1956-60), the relationship of body build and steroid levels (Tanner, Healy, Whitehouse, & Edgson, 1959), and physique and the performance of Olympic athletes (Tanner, 1964).

The work of Tanner, Ross (1978), and Malina (1980) in recent years has focused on the dynamics of structure and function. It is this dynamic relationship that draws a variety of disciplines together in the search for better understanding of the limits of human performance and the factors that set those limits. The so-called ecological rules of Bergmann (1847) and Allen (1877) had long been cited as explanations for human variability under climatic stress. These rules, which predict that the surface area/volume ratio would be low in cold climates where heat retention is important, could be alternatively interpreted as expression of the action of natural selection favoring the best adapted genotypes or as the result of environmental influence on the growth process. In either event the best-adapted phenotype would result. A number of human biologists have been intrigued by the relationships of body size and composition to the environment and have looked at populations of different racial and ethnic groups to identify the factors involved in determining the phenotype (Biasutti, 1959; Newman, 1961; Newman & Munro, 1955; Roberts, 1953, 1960). The evidence that cold and heat stress yield a specific anatomical response in humans is still inferential at best. This is partly because humans have been so skillful at modifying the environment to buffer themselves from the full impact of climatic factors.

Adaptation to High-Altitude Hypoxia

However, some environmental stressors are harder to buffer than others; among the most difficult is the form of hypoxia experienced at high altitude. Human biologists have sought indicators of the limits of human adaptability where populations have resided at altitudes sufficiently great to create hypoxic stress. The fact that millions of people live at altitudes greater than 3,000 meters has made it possible to measure the anatomical and physiological characteristics of high-altitude dwellers and compare them with low-altitude populations. Much of the work done at high altitude has been carried out in the Andes of South America, but significant studies have also been conducted in the Himalayas, in the Soviet Union, and in Ethiopia.

The literature on high-altitude adaptations is voluminous, and perhaps the best single source for the interested reader is the International Biological Programme volume, *The Biology of High-Altitude Peoples*, edited by Paul T. Baker (1978). From these studies, it appears that not all high-altitude populations adapt to hypoxic stress identically and that the adaptations seen in populations of long-term residence at high altitude are a mixture of genetically determined and developmentally acquired traits. The altitude thorax, the most striking aspect of the high-altitude dweller's anatomical adaptations, provides substantially greater lung capacity permitting effective aeration of tissues at a respiratory rate generally lower than that of sea-level dwellers. The large diffusing capacity of the high-altitude dweller's lungs is necessarily accompanied by an enlarged capillary bed and cardiovascular system with the capacity to circulate a greater volume of more viscous blood.

An important component of this adaptation is a tendency toward right ventricular hypertrophy in native dwellers at altitudes above 4,000 meters. The larger heart pumps a larger blood volume by increased stroke volume and a generally lower pulse rate. The obvious similarities between high-altitude adaptation and athletic conditioning have long interested exercise physiologists as well as other human biologists, and the nature of the structural-functional adjustments to chronic hypoxic stress is still being debated. With the movement of some high-altitude populations to low altitudes, the adaptations that allowed survival in their native habitat are no longer crucial. How will the growth and development processes of these populations respond to their new environment in future generations? From the information that is now accumulating, it appears that many of the anatomical characteristics associated with high-altitude adaptations are the product of altered growth patterns. The phenomenon of phenotypic plasticity responding to developmental acclimatization appears to yield a morphological configuration that very much resembles a genetic adaptation. This will be an important finding if it is ultimately borne out by the longer term observations of subsequent generations. The dynamic nature of structural/functional adjustments is perhaps nowhere more strikingly evident than in these populations.

The phasing of growth and maturation in Andean populations has been followed by Frisancho (1976), who noted that growth of the chest is accelerated while growth of the lower limbs is delayed. Ultimately, epiphyseal closures terminate growth in the lower limbs while the ratio of sitting height to stature is still high. The result is a short, barrel-chested body build that confers a number of advantages in the presence of chronic hypoxic stress. The mechanism by which this build is attained is thought to be selective stimulation and retardation of skeletal growth and maturation. Work performance of high altitude natives shows that developmental acclimatization reduces the physiological strain of performance in a hypoxic environment (Buskirk, 1976). Baker (1976) concludes that acclimatization to hypoxia occurring during adulthood does not improve work capacity to the degree seen in individuals growing up in a hypoxic environment.

While some attempts have been made to estimate the body composition of native high-altitude dwellers, interpretation of the results is complicated by the fact that populations living above 4,000 meters are subject to a number of stresses in addition to hypoxia. Cold stress is a frequent occurrence, and high winds, rough terrain, and high ultraviolet radiation levels are all part of the environment. Moreover, the quantity and variety of food is often severely restricted. In such a multistress environment, it is impossible to identify the effect of a single stressor. Animal experiments have been designed to control for hypoxia as the single variable. Where such experiments have been carried out, the effects of hypoxia generally parallel those observed in human populations with respect to the cardiovascular and respiratory systems, while the changes in skeletal anatomy are less clearly expressed. Because the mechanisms producing the characteristic skeletal proportions of the high-altitude dweller are thought to involve environmental influences on the growth process, it is possible that aspects of human growth are more susceptible to such influences than the growth processes of other species. Consideration of what those aspects might be will require a closer look at the peculiarities of the human growth curve and its evolutionary significance. This is the point where the anthropological focus on kinanthropometry provides a unique and, I hope, useful perspective on the relationships among human size, shape, and functional capacity.

Growth Curves of Mammals

Growth is a combination of *hyperplasia*, or increase in cell number, and *hypertrophy*, or increase in cell size. The earliest phases of growth, during which tissue differentiation and organogenesis are taking place, are characterized by predominance of hyperplasia. This roughly coincides with the embryonic stage of growth. The fetal stage increasingly involves hypertrophy. By the end of gestation, hypertrophy is the major form of growth in most tissues. In some tissues, as in the case of neurons, hypertrophy is the sole means of growth even before birth. In other tissues, such as epithelium and blood-forming tissues and liver, new cells are produced throughout life. All tissues do not grow at the same rate and the impact of environmental factors is generally greatest on tissues that are experiencing rapid growth at that time. Thus, prenatal damage may be limited to a single organ or sector of the anatomy. In postnatal life most stressors affect growth through a slowdown of hypertrophy, the effect being a symmetrical reduction in the rate of growth. Such allometric reductions in growth rate will produce an organism smaller than normal but with normal proportions. This is true for mammals in general and therefore for humans as well. Where humans differ from other mammals is in the length of time that elapses between birth and the attainment of sexual maturity. Humans are slow growers and late maturers and therefore experience an extended period wherein the environment can affect hypertrophic growth.

Besides growing slowly, humans experience spurts in growth that make their growth curve nonlinear. Examination of the human growth curve during prenatal life (see Figure 1) shows that maximal growth occurs around the time of midgestation. The postnatal growth curve (see Figures 2a, b; 3a, b) exhibits another kind of irregularity with a modest spurt just after birth and another spurt during adolescence. The adolescent growth spurt in both weight and height precedes the attainment of sexual maturity. In a series of comparative growth curves described by Scammon (1930), several organs exhibit a variety of patterns ranging from essentially linear, as in the case of the thyroid gland, to hyperbolic, as in the case of the thymus (see Figure 4). The episodic nature of so much of human growth makes the process vulnerable to environmental stressors in different areas and at different times. We are a long way from understanding just how much an individual's adult phenotype, health, athletic ability, behavior, and longevity are influenced by the unique patterns of oranismic-environmental interaction experienced over the first 15 to 20 years of postnatal life.

Why do humans take so long to complete the process of growth and maturation? This is a question of profound significance to the understanding of human biology, but one which must be answered speculatively. This is because a full understanding of the reasons for the uniqueness of human growth and development will require more information than is currently available concerning the evolutionary events that produced it. As it now stands, there are two lines of

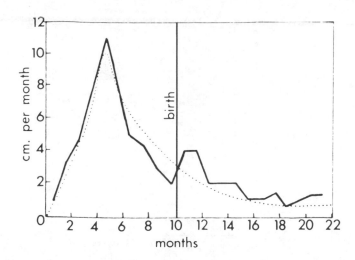

Figure 1. Velocity curve for stature before and after birth. Each point on the graph represents the mean increase in length during that particular month. *Note*. From *Growth and Form*, 2nd ed. (p. 112), by D'Arcy W. Thompson, 1942, London: Cambridge University Press. Copyright 1942 by Cambridge University Press. Reprinted by permission.

HUMAN GROWTH AFTER BIRTH

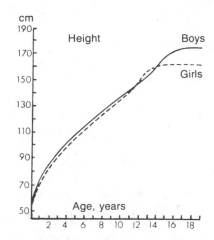

Figure 2a. Typical individual height-attained curves for English boys and girls. *Note*. From "Standards from birth to maturity for height, weight, height velocity, and weight velocity: British children, 1965, Part I" by J.M. Tanner, R.H. Whitehouse, and M. Takaishi, 1966, *Archives of Disease in Childhood*, **41**, p. 466. Copyright 1966 by British Medical Journal. Reprinted by permission of the authors and the editor.

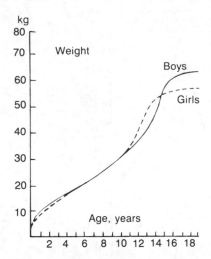

Figure 2b. Typical individual weight-attained curves for English boys and girls. *Note.* From "Standards from birth to maturity for height, weight, height velocity, and weight velocity: British children, 1965, Part I" by J.M. Tanner, R.H. Whitehouse, and M. Takaishi, 1966, *Archives of Disease in Childhood*, **41**, p. 466. Copyright 1966 by British Medical Journal. Reprinted by permission of the authors and the editor.

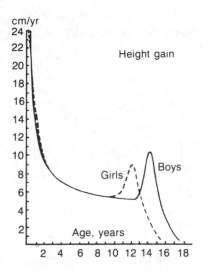

Figure 3a. Typical individual velocity curves for height: English boys and girls. *Note.* From "Standards from birth to maturity for height, weight, height velocity, and weight velocity: British children, 1965, Part I" by J.M. Tanner, R.H. Whitehouse, and M. Takaishi, *Archives of Disease in Childhood*, **41**, p. 467. Copyright 1966 by British Medical Journal. Reprinted by permission of the authors and the editor.

GROWTH IN HEIGHT AND WEIGHT

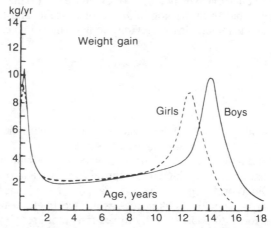

Figure 3b. Typical individual velocity curves for weight: English boys and girls. Note. From "Standards from birth to maturity for height, weight, height velocity, and weight velocity: British children, 1965, Part I" by J.M. Tanner, R.H. Whitehouse, and M. Takaishi, 1966, *Archives of Disease in Childhood*, **41**, p. 467. Copyright 1966 by British Medical Journal. Reprinted by permission of the authors and the editor.

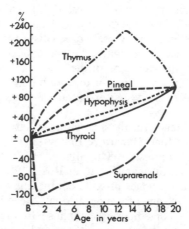

Figure 4. Growth of certain internal organs. The lower three are endocrine glands; the functions of the other two are not so certain. The curves show percentage of adult weight attained at a given time. *Note.* From "Measurement of the body in childhood" (p. 200) by R.E. Scammon, 1930, in J.A. Harris, C.M. Jackson, D.G. Paterson, and R.E. Scammon (Eds.), *The Measurement of Man*, Minneapolis, University of Minnesota: University Press. Copyright 1930 by University of Minnesota Press. Reprinted by permission.

evidence bearing on the question, but the validity of assumptions that they are linked is open to question. The two lines of evidence are (a) the fossil evidence for evolutionary change in the hominid line of descent, and (b) the growth and anatomical characteristics of living nonhuman primates such as the chimpanzee.

Fossil Evidence

During the past 30 years, substantial fossil evidence of the history of our species has been recovered. Most of this material has been found in Africa, but there have been important discoveries in Asia as well, giving support to the belief by many paleontologists that human evolution took place across a broad expanse of the Old World tropics. The materials that have been found indicate a progression from a small, apelike, tree-dwelling primate to a slightly larger semiterrestrial form, to a larger, fully terrestrial semi-erect biped, to a fully erect biped that underwent additional evolutionary change, thus giving rise to *Homo sapiens* of contemporary form perhaps as much as 50,000 years ago. Although there are numerous differences of opinion among human paleontologists concerning the sequence of events and the dates of certain threshold events, the fossil evidence is sufficient to reconstruct most of the important phases of the evolutionary progression leading up to our species.

In the fossil record, certain major trends are well documented. One such trend is the progressive change in skeletal morphology that culminated in the attainment of fully erect posture (Lovejoy, 1978). This is a crucial aspect of contemporary human behavior and many students of human evolution would argue that the attainment of fully erect bipedalism was the key event in the origin of our species (Rose, 1984a). The anatomical concomitants of erect bipedalism, including the balancing of the skull on the top of the spinal column, made possible the expansion of the brain and the reorientation of the pharynx, both complementary and essential elements for the emergence of speech.

The anatomical evidence for the attainment of erect bipedalism is seen in fossilized bones of the foot (Oxnard & Lisowski, 1980), lower limb, pelvis (McHenry, 1975a, 1975b; Rose, 1984b; Sigmon, 1982; Stern & Susman, 1983; Susman & Stern, 1982; Susman, Stern, & Rose, 1983), spinal column, and skull. In aggregate, these bones provide compelling evidence of the change in posture from the semi-erect form occasionally seen in some contemporary apes to that seen in contemporary humans. In order to appreciate the significance of these postural changes, it is useful to examine the anatomy of our closest living relatives, the great apes.

Anatomy and Posture of Contemporary Apes

On the basis of biochemical similarities, chromosome configurations, and a variety of anatomical characteristics, the closest living relatives of our species are to be found among the great apes. Of these, most biologists favor the chimpanzees of Africa as the closest relative. However, the gorilla of Africa and the orangutan of Southeast Asia are believed by some to be more closely related to *Homo sapiens*. On the basis of anatomical characteristics, all three of these primate species share certain important adaptations. One such adaptation is the ability to brachiate, or swing through tree branches while suspended by the forelimb. While the ability to brachiate is not often exploited by the gorilla, it is used on occasion, and the relationships of the scapula, clavicle, humerus, radius, ulna, carpals, metacarpals, and phalanges that make it possible form an important, integrated complex that closely resembles that seen in chimpanzees and orangutans. Associated with the definitive characteristics of the brachiator's shoulder and forelimb are changes in the spine that shorten and stiffen the back. Altogether, the upper body of the brachiating ape is unique in the skeletal morphology necessary for locomotion in the suspensory mode. In this respect they differ significantly from the other nonhuman primates who either behave as quadrupeds while in trees, or, as in the spider monkey of South America, use a prehensile tail as a suspensory fifth limb.

The anatomy of the brachiating apes bears many striking resemblances to our own in the shape and flexibility of the shoulder joint and in the shortened and stiffened spine with the associated arrangement of the viscera in a broad anterio-posteriorly flattened thorax. The similarities between the human and ape upper body are such that a common ancestor seems likely. However, major anatomical differences are found in the traits associated with the attainment of erect bipedalism. Thus, changes in the position and shape of the skull, the pelvis, and the lower limb appear to have occurred in the human line of descent after divergence from that of the apes.

Other aspects of the biology of contemporary apes are of interest in the understanding of the human pattern of development. The tree-dwelling brachiating apes do not have large litters of young. In fact, single births predominate. The reason for this is obvious. Care of a single infant, or at most two, is virtually all that can be managed by a mother who must move about in trees. The limit that the arboreal habitat places on litter size has the beneficial side effect of allowing the mother to concentrate all her attention on a single infant. The result is a great deal of intensive individual care for the infant, whose early months and years are filled with intense socialization by its mother. With a guarantee of continual maternal care, the infant can be born in a relatively immature state and can be permitted an extended period of dependency while being programmed for a complex and highly socialized adult life. Somewhere in primate history, the combination of prolonged dependency and an active and complex social pattern was formed; the higher primates (monkeys and apes) that have survived into modern times all exhibit some permutation of social grouping, along with a great deal of flexibility in its expression.

Humans have inherited the tendency toward complex socialization patterns that characterize all higher primates and have enlarged upon it. The complexity and variability of human organization is substantially greater than that seen in any other primate, but its roots are discernible in the prevailing primate pattern. One of the major mechanisms through which humans have achieved a unique adaptation has been through the expansion and intensification of the learning experience inherent in a prolonged period of dependency. Merely prolonging the period of dependency would not, however, be sufficient to produce the behavioral differences we see between contemporary humans and nonhuman primates. While the provision of additional time to write elaborate software is essential, without sophisticated hardware to run it not much would be accomplished. The coincidence of slow maturation and a repositioning of the skull that permitted expansion of the brain was an event that brought the hardware and the software together. Slow maturation was part of the primate heritage, but repositioning of the skull with the attainment of erect bipedalism was uniquely human. The advantages accruing to the possessors of the combination have been sufficiently great to offset the obvious disadvantages.

Growth Spurts

The weak, uncoordinated, highly vulnerable state in which the human infant is born creates enormous demands on the mother. Moreover, these demands persist for many months. Even after the human infant achieves sufficient neurological organization to walk on two feet, it may take another 15 years to attain sexual maturity. The period of time from birth to sexual maturation is a time when both physical and intellectual growth are taking place. In humans, intellectual growth may continue to the end of life, even though neurological growth is completed before the age of 20. One reason for the great learning potential humans possess is the complexity of the interactions between neurons that is attained during their extended period of dependency.

The uniqueness of the human growth curve is one aspect of the mode of attainment of the characteristics that we consider quintessentially human: learning, communication, and innovativeness. Inspection of the human growth curve reveals that there are two periods of growth acceleration that give the curve a distinctly nonlinear shape. There are hints of this nonlinearity in the growth curves of other primates, especially the chimpanzee, but only in humans do we encounter the intense acceleration seen during the adolescent growth spurt. Seen as a distance curve, the pattern is quite clear, but its true dimensions are best appreciated when viewed as a velocity curve (see Figures 3a and 3b). The velocity curve approach allows focus on the increments of growth during a specific period of time, allowing such increments to be treated as "vector values." Viewed as vector values, the increments occurring during the brief period immediately following birth and the months of the adolescent growth spurt are seen to be substantial.

Catch-Up Growth

The spurt of growth following the first week postpartum has the characteristics of what is commonly called "catch-up growth." Catch-up growth is frequently seen after some growth-retarding event, such as malnutrition, severe illness, or emotional disturbance. When sufficient nourishment is available and the stressor is removed, growth acceleration occurs for a period of time sufficient to place the individual's attained growth status back on the normal growth curve. Catch-up growth can compensate for growth retardation if retardation is not excessively prolonged, but it may not fully compensate for extended periods of growth arrest. In the period following birth, the human infant compensates for the slowed growth it experiences during the last weeks of gestation when the diffusing capacity of the placenta is challenged and when intrauterine space is becoming inadequate.

The brain of the human fetus has become exceptionally large in comparison with the rest of the body by the last month of gestation. Its growth and organization enjoy the highest priority in the internal distribution of resources available to the fetus. Its growth is maintained at the expense of linear growth at the end of gestation, and it continues to grow rapidly for the first year after birth. The processes of myelination and dendritic arborization account for most brain growth during this time, with no addition of neurons (Winick, 1976; Winick, Brasel, & Rosso, 1972). The speed and complexity of impulse transmission that characterize the human brain are being potentiated at this time. The end product of this phase of human growth is the hardware that will run the programs we call learning. After the initial spurt of catch-up growth, the physical growth of the human child settles down to a slow, steady pace that is essentially linear. This linear phase of growth ends at the onset of the adolescent growth spurt when the human growth curve is marked by rapid increments in both weight and height under the influence of sex hormones.

Adolescent Growth Spurt

The adolescent growth spurt is a time of change in body composition accompanying increases in weight and stature. As it nears completion, males will have become more muscular while females will have increased in adiposity. The end of the adolescent growth spurt is followed by a variable period of slow incremental growth as the epiphyses of the long bones and finally those of the vertebrae close. There is a modest adolescent growth spurt in chimpanzees and possibly in macaques, but it occurs much earlier than in humans (Watts & Gavan, 1982). In male humans, chimpanzees, and macaques, the spurt of linear growth follows maturation of the testis and the beginning of spermatogenesis. This male pattern is late in the maturation process as opposed to the pattern in human and chimpanzee females where the adolescent growth spurt is close to the age of menarche.

One interpretation of the adolescent growth spurt is that it represents another example of catch-up growth. The deceleration in the rate of growth that persists throughout childhood is a period when a child can learn to think and behave like an adult without being forced to compete for sexual activity. It is a time when situations can be simulated and ultimately anticipated without the risks of real-life consequences. When the neurological system has had sufficient time to achieve its full potential, the constraints that had held growth to a low level are relaxed, and a period of catch-up follows. Thus, the pattern that is so pronounced in humans reflects the unique dependence on learning and socialization that humans have developed during the evolution of the species. The intimations of this pattern as seen in chimpanzees and macaques allow the possibility that a tendency toward the dependence existed relatively early in the evolutionary history of primates, at a time before the ancestral lines of Old World monkeys, apes, and humans diverged. With the increase in human brain size made possible by the attainment of erect bipedalism, the human and ape lines of descent diverged, with the human commitment to elaboration of social-cultural adaptation becoming the predominant element of subsequent human evolution.

Sexual Differences in Body Composition and the Timing of Growth

Sexual maturation in humans results in the maximization of sexual dimorphism seen in the young adult. Body composition in males exhibits the highest percentage of skeletal muscle at this time, while the composition of the female body has shifted toward greater adiposity as reserves essential for successful reproduction are established. Females, on the average, are about 2 years ahead of males in the state of maturation when growth stops. Interestingly, this female lead in maturation can be traced back to the 20th intrauterine week when the female fetus is about 10% ahead of the male in appearance of ossification centers and other indicators of maturation. Thus, although generally somewhat smaller than male newborns, females are more mature and simultaneously more viable from the beginning of life. At the time of conception, males outnumber females in our species; the primary sex ratio (ratio of male to female zygotes at the time of fertilization) is estimated to be 136/100. By the time of birth, the ratio (the secondary sex ratio) in the United States has declined to 106/100. The ratio has reached 100/100 at about 18 years of age, and presently is about 100/138 at age 65 and 100/156 among individuals living longer than 75 years.

During the childhood years there is very little difference between male and female body composition or body size. For a time, when the female advancement in maturation leads to expression of the adolescent growth spurt, females may be somewhat larger than their male classmates. When the females have completed growth, most of the males their age will still be growing and will finally exceed them in weight and stature (Tanner, 1969).

In both sexes, the growth process produces substantial alterations in proportionality. At birth, the human infant's head accounts for about 25% of its total length. In the adult, the head accounts for about 10% of total stature. As seen in Table 1, the greatest gain in size is made by the lower limbs, which increase from 15% of total length at birth to 30% in the adult. The proportions of the trunk and upper limbs change very little during the period from growth to maturity. In his analysis of the changes in the center of mass in the growing macaque, Grand (1981) saw a similar, but less pronounced shift in the relative sizes of the head and lower limb.

Over the years, a number of comparative studies of the segmental distribution of body mass have been undertaken (Grand, 1977a, 1977b, 1978, 1981, 1983). Changes in the center of mass have been shown to be a general phenomenon, but the location of the center of mass differs in accordance with differences in posture and locomotion. Zihlman and Brunker (1979) applied a similar technique to adult chimpanzees and compared their values to those of adult humans. In the chimpanzees the lower limbs make up 18% of the total body weight, the trunk 66.5%, and the upper limbs 15.5%. In humans,

Table 1. Relative sizes[a] of parts of the body

Age	Head and neck	Trunk	Upper limbs	Lower limbs
Birth	30	45	10	15
2 years	20	50	10	20
6 years	15	50	10	25
Adult	10	50	10	30

[a]Approximate percentages of total body volume.

the lower limbs make up 32%, the trunk 58.6%, and the upper limbs 9.4% of total body weight. Clearly, the attainment of erect bipedalism has led to a major alteration in the relative size of the lower limb in humans. The magnitude of the difference between the chimpanzee and human lower limb is even more pronounced when the anatomy of the foot and pelvis is examined. Additional evidence of major evolutionary change in the lower limb of humans is found in the embryological relationships of muscles and nerves in the human thigh where rotation has resulted in a reversal of the flexor/extensor innervation of the thigh.

Attainment of Erect Posture

Human erect bipedalism requires the development of several spinal curvatures to enable the center of mass to be placed at the point where weight is transferred from the trunk to the lower limb. At birth, human infants have two primary curvatures of the spine, both anteriorly convex: one in the thoracic area and the other in the sacrum. The sacral curvature later becomes fixed by the fusion of the sacral vertebrae. The thoracic curve remains somewhat flexible throughout life, although the intervertebral disks in the thoracic region are relatively thick, and movement is limited by the presence of overlapping laminae and the oblique vertebral spines.

When the human infant begins to hold its head up while in the prone position, usually at about 3 months of age, a secondary curvature appears in the cervical spine. This curvature is convex anteriorly and will remain mobile throughout life, with its radius determined by the tension of muscles crossing its convexity. The intervertebral disks in the cervical area are thick and allow considerable flexibility in movement. The adult human skull is balanced on top of this flexible cervical spine, but this is not the case in the newborn, where the angle of articulation resembles that of monkeys and apes. With growth and maturation, the human skull grows both forward and backward from the foramen magnum in roughly equal amounts, making a balanced load on top of the spine possible. In the other anthropoids, however, growth in the anterior aspects of the skull is much greater than in the posterior aspects, leading to much greater mass of the skull in front of the spine. This relationship is harmonious, with the suspension of the skull in front of the body supported by nuchal musculature while in a quadrupedal or semiquadrupedal position. This is a major difference between humans and apes. At the time the human infant begins to sit up, one more curve in the lumbar region appears. The curvature is also convex forward and will remain mobile, being shaped by a combination of intervertebral disks and vertebral musculature (Sinclair, 1973).

The first efforts at locomotion by the human infant are usually in a quadrupedal mode or some variation thereof. Early attempts to assume a bipedal posture are generally awkward and unsuccessful for several reasons. First, the center of mass is still too high at this stage of development, and the pelvis must be forced to grow outward to allow weight transfer to occur at the acetabulum. As more time is spent upright, the weight of the trunk pressing down on the sacrum and being transmitted to the auricular surfaces of the pelvis will reshape the pelvic basin and enhance stability in the upright position. This all takes time and repeated effort, and usually does not even begin until the infant is nearly 1 year old. However, when it has been accomplished, the unique gait that separates humans from other primates gives the young human the capacity to walk great distances, sprint at high speed, or run a marathon.

The phylogeny and ontogeny of human erect bipedalism are of great interest in their own right. When coupled with the important role erect posture has played in allowing the human brain to expand as it is balanced on top of the spine without the constraints of a massive nuchal musculature to suspend it, the way we stand, walk, and run emerges as a human trait complex of central importance in shaping our major adaptive strategy. Although it is often said that humans are weak and vulnerable animals who are dependent upon their culture for survival, we are, in reality, admirably equipped to compete in a variety of habitats. Our variable gait, ability to cool ourselves by evaporation, omnivorous dietary capacity, and ability to remain active during daylight and night hours make our species formidable by any standard.

Body Composition and Nutrient Reserves

Human body composition includes a combination of muscle, bone, and fat that can be drawn upon for the maintenance of metabolic needs when food is scarce. Even formidable hunter-gatherers experience times of scarcity and must be able to survive by metabolizing endogenous reserves. A wounded hunter may not be able to pursue game. At the same time that the demands of wound-healing are being felt, the availability of protein is curtailed. By a shift in insulin sensitivity, muscles stop incorporating amino acids even while they continue to release them into the vascular system. In this manner, uninjured skeletal muscle helps to sustain the healing process in injured tissues and supply amino acids necessary to support the immune system in fighting infection. Likewise, bone is the ultimate reservoir for calcium that is essential for the regulation of muscle contraction. Through an intricate system of endocrine responses, adjustments are made in the level of serum calcium to maintain physiological levels of 7 to 13 mg%. Fat is consumed to supply energy needs when serum glucose and glycogen stores are depleted. Amino acids from muscle can be consumed for energy by gluconeogenesis. These and other forms of endogenous reserves allow humans to range over great distances in pursuit of food, an ability that must certainly have played an important role in the expansion of human populations over most of the earth's surface. Kinanthropometrists are interested in the relationship between form and function and in the interaction of anatomy, growth, and performance. Although this is generally viewed as the province of exercise physiologists, anatomists, and athletic coaches and trainers, it is an area of broad interest to biologists who are concerned with the phenomenon of adaptation.

When one considers the basic mechanisms of metabolism as a means of supplying the action requirements of the organism, the limits of performance take on theoretical as well as practical significance. Is there an ideal anatomical configuration for performing a given task? Selective breeding of domestic animals supports the argument that superior performance can be bred. Thoroughbred horses and fighting bulldogs are two examples of highly specialized animals that excel in certain forms of performance. Human athletes are selected through motivation and training to maximize their capabilities in highly specialized events. Champion gymnasts are anatomically distinct from champion weightlifters. Sprinters are generally anatomically different from long-distance runners, and swimmers differ from pole vaulters. Perhaps the most interesting thing about humans is the fact that we can select so many athletic specialists out of a stock that has a common adaptive strategy. From the standpoint of the anthropologist, this is a matter of considerable theoretical interest.

Human Adaptive Strategy

The human adaptive strategy has capitalized on certain major evolutionary events to produce a highly flexible and exceptionally effective response system (Stini, 1979). With the development of a highly complex culture, the environment itself has been brought under control. Where cultural means are not effective in modifying the environment, there are a variety of behavioral adjustments that can be invoked to modify stress and to avoid strain. Such behavioral adjustments can involve small-group cooperation (necessarily involving the ability to communicate) or individual strategies, still dependent upon the ability to learn and apply the proper responses. When behavioral adjustments are insufficient to deal with a stressor, a series of physiological responses come into play. Short-term physiological modifications, often referred to as acclimatizations, can prevent damage without committing the organism to more profound changes that could prove maladaptive when the environment changes. Habituations can deal with stressors through accommodation without invoking responses that could in themselves be harmful. Upon sustained exposure, stimuli that are highly irritating can become insensible without damage to the organism. Such habituations are commonly seen in areas of sustained cold stress. Acclimatizations can be reversible, as in the case of seasonal acclimatizations where a number of physiological systems are adjusted to reduce the stress of seasonal change.

Other acclimatizations, as discussed earlier in the context of high-altitude physiology, can become permanently fixed in the anatomy and physiology of the individual. As in the case of the Andean high-altitude dweller, it is often difficult to differentiate developmental acclimatizations from genetically-determined adaptations. When populations living in areas where sustained stresses are experienced throughout the period of growth and devlopment over a period of generations, certain traits may appear to be genetic adaptations resulting from natural selection. In many cases, the only way to determine whether the adaptation is genetic or ontogenetic is to watch the results of a generation living in the absence of the stressor. We are currently studying such natural experiments in populations of high-altitude dwellers who have migrated to low altitudes in South America.

On a worldwide scale, the secular trend in increased body size and earlier sexual maturation has provided evidence that environmental factors until now had prevented large numbers of people over many generations from achieving their genetically determined growth potential (Damon, 1965; Frisancho, Sanchez, Pallardel, & Yanez, 1973). Thus, the smaller body size of the Japanese population prior to 1950 could be viewed as a developed acclimatization to a diet low in protein and calories. The rapid increase in stature and weight in the contemporary Japanese population and the slower increase that has taken place in the populations of North America and Europe over the past 100 years might look very much like the replacement of one population by another to a naive human paleontologist at some future date. In reality, it is an excellent example of the degree to which humans can reshape themselves to better adapt to their environment. In the human adaptive strategy, developmental acclimatization is the most profound change other than elimination of genes through natural selection.

Humans place a high value on the individual. It is interesting that human ethics are so concerned with the well-being, the rights, and the preservation of the individual. This is because our adaptive strategy has placed a high biological value on the survival of the individual. We are primates who have primarily single births after a lengthy gestation period. As erect bipeds with highly complex neurological systems to develop, we are slow maturers and begin reproductive life very late compared to other species. Along with slow growth and late maturation, we have long lives, and, most unusually, we have long postreproductive lives. In the human female, menopause occurs around the age of 50 years. In the United States today, life expectancy of a newborn female is over 78 years (U.S. Bureau of the Census, 1983). What is the significance of this lengthy period after childbearing is no longer possible? This is a question for which there is no ready biological answer. However, if we view our species as one which has benefited by the presence of older individuals through the elaboration and transmission of culture, survival of older, non-reproducing individuals does make sense. It also provides convincing evidence that in our species, culture is a part of our biology. Just as the human infant struggles to become an erect biped, to learn a language, and to communicate with other humans, adult humans are intrigued and attracted by other humans. This attraction has developed into a reverence for human life and a commitment to its preservation whenever possible. Although these principles of human behavior have often been violated, they are values that have been professed by civilizations throughout history and probably long before.

Conclusion

In the present context, the meaning of this view of human biology is that it has permitted our species to populate this entire planet with all of its climatic variation. Human variation has been maintained through the ability to adjust to the prevailing conditions without becoming genetically isolated from other human populations. The preservation of the integrity of our species has been further aided by the human tendency to seek out and interact with other human populations. As a result, our species is extremely polymorphic and polytypic, while at the same time fully interfertile. From the standpoint of the kinanthropometrist, whose interest in the interaction of human form and function creates an acute awareness and appreciation of human variation, it is essential to know the roots of this variation. In the training of athletes, we see individuals strive to achieve the maximum performance possible within anatomical and physiological limits. In the range of athletic events humans are able to perform, we are a species without equal. The preservation of genetic variability combined with our lengthy period of trainability makes the creation of specialists in many different areas of activity an intrinsic element of our biological adaptation. This is where the interests of the anthropologist and the kinanthropometrist converge.

References

Allen, J.A. (1877). The influence of physical conditions in the geneses of species. *Radical Review*, **1**, 108-140.

Baker, P.T. (1976). Work performance of highland natives. In P.T. Baker & M.A. Little (Eds.), *Man in the Andes* (pp. 300-314). Stroudsburg, PA: Dowden, Hutchinson and Ross.

Baker, P.T. (Ed.). (1978). *The biology of high-altitude peoples*. Cambridge: Cambridge University Press.

Beneke, F. (1878). *Anatomische grundlagen der constitutionsanomalien des Menschen*. Marburg.

Beneke, F. (1881). *Constitution und constitutionelles kranksein des menschen*. Marburg.

Bergmann, C. (1847). Uber die verhältnisse der Wärmeökonomie der thierezuihrer Grösee. *Göttingen Studien*, **3.5**, 95-108.

Biasutti, R. (1959). *Lerazzee ipopoli della terra*. Torino: UTET.

Brozek, J. (Ed.). (1965). *Human body composition. Symposia of the Society for the Study of Human Biology* 7. London: Pergamon.

Buskirk, E.R. (1976). Work performance of newcomers to the Peruvian highlands. In P.T. Baker & M.A. Little (Eds.), *Man in the Andes* (pp. 283-299). Stroudsburg, PA: Dowden, Hutchinson and Ross.

Damon, A. (1965). Stature increase among Italian-Americans; environmental, genetic or both? *American Journal of Physical Anthropology*, **23**, 401-408.

Frisancho, A.R. (1976). Growth and morphology at high altitude. In P.T. Baker & M.A. Little (Eds.), *Man in the Andes* (pp. 180-207). Stroudsburg, PA: Dowden, Hutchinson and Ross.

Frisancho, A.R., Sanchez, P.D., & Yanez, L. (1973, April). *Adaptive significance of small body size under poor socioeconomic conditions in Southern Peru*. Paper delivered at the annual meeting of the American Association of Physical Anthropology, Dallas, TX.

Grand, T.I. (1977a). Body weight: Its relation to tissue composition, segment distribution, and motor function. *American Journal of Physical Anthropology, 47*, 211-240.

Grand, T.I. (1977b). Body weight: Its relation to tissue composition, segment distribution, and motor function, II. Development of *Macaca mulatta. American Journal of Physical Anthropology.* **47**, 241-248.

Grand, T.I. (1978). Adaptations of tissue and limb segments to facilitate moving and feeding in arboreal folivores. In G.G. Montgomery (Ed.), *The Ecology of Arboreal Folivores* (pp. 231-241). Washington: Smithsonian Institution Press.

Grand, T.I. (1981). The anatomy of growth and its relation to locomotor capacity in *Macaca.* In A.B. Chiarelli & R.S. Corruccini (Eds.), *Primate evolutionary biology.* New York: Springer-Verlag.

Grand, T.I. (1983). Body weight: Its relationship to tissue composition, segmental distribution of mass, and motor function, III. The *Didelphidae* of French Guyana. *Australian Journal of Zoology, 31*, 299-312.

Heath, B.H., & Carter, L. (1967). A modified somatotype method. *American Journal of Physical Anthropology, 27*, 57-74.

Kretschmer, E. (1921). *Korperbau und Charakter.* Berlin: Springer-Verlag.

Lovejoy, C.O. (1978). A biomechanical review of the locomotor diversity of early hominids. In C.J. Jolly (Ed.), *Early hominids of Africa* (pp. 403-429). New York: St. Martin's Press.

Malina, R.M. (1980). Physical activity, growth, and functional capacity. In F.E. Johnston, A.F. Roche, & C. Susanne (Eds.), *Human physical growth and maturation* (pp. 303-327). New York: Plenum Press.

McHenry, H.M. (1975a). A new pelvic fragment from Swartkrans and the relationship between the robust and gracile australopithecines. *American Journal of Physical Anthropology, 43*, 245-262.

McHenry, H.M. (1975b). Multivariate analysis of early hominid pelvic bones. *American Journal of Physical Anthropology, 43*, 263-270.

Newman, M. (1961). Biological adaptation of man to his environment: Heat, cold, altitude and nutrition. *Annals of the New York Academy of Sciences, 91*, 617-633.

Newman, R.W., & Munro, E.H. (1955). The relationship of climate and body size in U.S. males. *American Journal of Physical Anthropology, 13*, 1-17.

Oxnard, G.E., & Lisowski, F.P. (1980). Functional articulation of some hominid foot bones: Implications for the *Oldovai* (Hominid 8) foot. *American Journal of Physical Anthropology, 52*, 107-117.

Roberts, D.F. (1953). Body weight, race and climate. *American Journal of Physical Anthropology, 11*, 533-558.

Roberts, D.F. (1960). Effects of race and climate on human growth as exemplified by studies on African children. In J.M. Tanner (Ed.), *Human growth* (pp. 59-72). New York: Pergamon Press.

Rose, M.D. (1984a). Food acquisition and the evolution of positional behavior: The case for bipedalism. In D.J. Chivers, B.A. Wood, & A. Bilsborough (Eds.), *Food acquisition and processing in primates.* New York: Plenum Press.

Rose, M.D. (1984b). A hominine hip bone, KNM-ER 3228, from East Lake, Turkana, Kenya. *American Journal of Physical Anthropology, 63/4*, 371-378.

Ross, W.D. (1978). Kinanthropometry: An emerging scientific technology. In F. Landry & W.A.R. Orban (Eds.), *Biomechanics of sports and kinanthropometry* (p. 269). Miami: Symposia Specialists.

Scammon, R.E. (1930). The measurement of the body in childhood. In J.A. Harris, C.M. Jackson, D.G. Paterson, & R.E. Scammon (Eds.), *The measurement of man.* Minneapolis: The University of Minnesota Press.

Sheldon, W.H., Dupertuis, C.W., & McDermott, E. (1954). *Atlas of men*. New York: Harpers.

Sheldon, W.H., Hartl, E.M., & McDermott, E. (1949). *The varieties of delinquent youth*. New York: Harpers.

Sheldon, W.H., & Stevens, S.S. (1942). *The varieties of temperament*. New York: Harpers.

Sheldon, W.H., Stevens, S.S., & Tucker, W.B. (1940). *The varieties of human physique*. New York: Harpers.

Sigmon, B.A. (1982). *Comparative morphology of the locomotor skeleton of Homo erectus and the other fossil hominids, with special reference to the tautavel inominate and femora*. Paper presented at the 1er Congres International de Paleontologie Humaine, Nice, France.

Sinclair, D. (1973). *Human growth after birth*. London: Oxford University Press.

Stern, J.T., & Susman, R.L. (1983). Locomotor anatomy of *Australopithecus afarensis*. *American Journal of Physical Anthropology, 60*, 279-317.

Stini, W.A. (1979). Adaptive strategies of human populations under nutritional stress. In W.A. Stini (Ed.), *Physiological and morphological adaptation and evolution* (pp. 387-407). The Hague: Mouton Publishers.

Susman, R.L., & Stern, J.T. (1982). Functional morphology of *Homo habilis*. *Science, 217*, 931-934.

Susman, R.L., Stern, J.T., & Rose, M.D. (1983). Morphology of KNM-ER 3228 and O.H. 28 inominates from East Africa. *American Journal of Physical Anthropology, 60*, 259.

Tanner, J.M. (1956). Physique, character and disease: A contemporary appraisal. *Lancet, 2*, 635-637.

Tanner, J.M. (1963). The regulation of human growth. *Child Development, 34*, 817.

Tanner, J.M. (1964). *The physique of the Olympic athlete*. London: George Allen and Unwin.

Tanner, J.M. (1969). *Growth at adolescence*. Oxford: Blackwell Scientific Publishers.

Tanner, J.M., Healy, M.J.R., Whitehouse, R.H., & Edgson, A.C. (1959). The relation of body build to the excretion of 17-ketosteroids and 17-ketogenic steroids in healthy young men. *Journal of Endocrinology, 19*, 87-101.

Tanner, J.M., & Inhelder, B. (Eds.). (1956-60). *Discussions on child development* (Vols. I-IV). London: Tavistock Publications.

U.S. Bureau of the Census. (1983). *Americans in transition: An aging society* (Current population reports, series P-23, No. 128). Washington, DC: U.S. Government Printing Office.

Viola, G. (1935). Criteres d'appreciation de la valeur physique, morphologique et fonctionelle des individus [Standards for assessment of individual status: Physical, morphological, and functional]. *Biotypologie, 3*, 93.

Viola, G. (1936). Il mio metodo di valutazione della constituzione individual [My method for assessment of individual physical status]. *Endocrinol Patol. constit., 12*, 387.

Watts, E.S., & Gavan, J.A. (1982). Postnatal growth of nonhuman primates: The problem of the adolescent spurt. *Human Biology, 54*(1), 53-70.

Winick, M. (1976). *Malnutrition and brain development*. New York: Oxford University Press.

Winick, M., Brasel, J.A., & Rosso, R. (1972). Nutrition and cell growth. In M. Winick (Ed.), *Nutrition and development* (pp. 59-67). New York: Wiley.

Zihlman, A., & Brunker, L. (1979). Hominid bipedalism: Then and now. *Yearbook of Physical Anthropology, 22*, 132-162.

PART II

A Maturational Perspective

The theme of Part II is the relationship between the maturational status and physical performance of children and adolescents. The focus of Robert Malina's keynote paper is on young athletes, defined by their performance levels and degree of commitment to their sporting pursuits. In contrast, the papers by Bastos and Hegg and by Hebbelinck and his colleagues concern themselves with unselected groups of school children.

Malina's paper is an extensive exploration of what is known and suspected about the maturation of young athletes. Malina's research and his familiarity with the literature contribute to the insightful synthesis he presents. He examines the strengths and weaknesses of a number of maturity indicators, taking time to chide us for being unduly interested in young female athletes because one of their indicators (menarche) is so readily available. He infers that although training-maturity associations in males may be similar to those in females, we know much less about these associations in males because the information is harder to obtain.

Bastos and Hegg's article presents cross-sectional data on Brazilian boys 10 to 17 years of age. Their findings, which indicate a strong relationship between strength and sexual maturity, also serve to confirm the inadequacy of chronological age as an index of maturity status.

In their paper, Hebbelinck and his colleagues use data from the ongoing Leuven Growth Study to examine the relationships between skeletal maturity and a number of fitness and performance variables among children 6 years of age. The Belgian early maturers performed significantly better than the late maturers on just two of eight test items, both of which were strength-related.

2

Maturational Considerations in Elite Young Athletes

Robert M. Malina
UNIVERSITY OF TEXAS
AUSTIN, TEXAS, USA

Any consideration of maturity of elite young athletes is immediately faced with several problems. The first is a matter of definition. What is an athlete? Young athletes are usually defined by success in club and/or age-group competitions, on interscholastic or agency teams, and in provincial, national, and international competitions. Thus, it is necessary to carefully define the samples being described and compared.

Selection is a second problem when considering young athletes. Successful young athletes are a highly selected group, selected primarily on the basis of skill, but sometimes on the basis of size and physique in some sports or positions within a sport. Selection may be by oneself, by parents and coaches, or by both. Self-selection is indeed a critical factor, for example, a youngster's motivation to train and to be receptive to coaching. Selection also occurs to some extent by default, that is, those dropouts who chose not to participate for reasons of ability or reasons unrelated to sport. As competition becomes more rigorous, selection is probably a more important factor.

A third problem deals with variation in maturity because successful young athletes in many sports often differ in maturity status compared to nonathletes. Maturity and growth differ fundamentally. Maturity implies progress toward the mature state, which varies with the biological system considered. Skeletal maturity is a fully ossified skeleton; sexual maturity is reproductive capability; dental maturity is the eruption of a full complement of permanent teeth. Growth, on the other hand, implies changes in size. All individuals end up as adults skeletally mature or sexually mature, but all end up as adults with different heights. This is the fundamental distinction: progress towards maturity versus ultimate size. Both processes are probably under separate genetic regulation; yet they are related (Tanner, 1962). Youngsters advanced in maturity differ in size, physique, body composition, and performance compared to those delayed in maturity (Malina, 1984). The differences between contrasting maturity groups are especially pronounced during adolescence which includes the age span from about 9 through 17 years of age.

This chapter addresses several issues related to maturity in young athletes: (a) the assessment of maturity and the relationships among maturity indicators; (b) the effects of regular training on maturity indicators; and (c) implications for young athletes in several sports.

Interrelationships Among Maturity Indicators

The most commonly used indicators of maturity are skeletal maturation, the development of secondary sex characteristics, the timing of menarche, and the timing of maximum growth in height (peak height velocity). Dental maturation is another occasionally used indicator. Two questions are of importance in considering different maturity indicators. First, do the indicators measure the same kind of biological maturity? That is, if a youngster is rated as maturationally advanced by one indicator, is he or she rated as advanced by another? With the exception of dental maturity, which tends to proceed independently, the other indicators of maturity are positively related (Bielicki, Koniarek, & Malina, 1984; Demirjian, 1978; Malina, 1978; Tanner, 1962). Second, are ratings of maturity consistent over time? Is a youngster maturationally delayed at 6 years of age also delayed at 11 years of age? The answer is generally yes, although there may be variation during adolescence.

Issues of methodology are also of concern. Secondary sex characteristics are useful only during adolescence and are ordinarily rated on 5-stage scales for each trait. Yet, the development of these characteristics is a continuous process upon which arbitrary discontinuities are imposed. The process of skeletal maturation is the same in all individuals and covers the entire developmental period. Yet, methods of estimating skeletal maturity vary. For example, are those judged advanced or delayed by the Greulich and Pyle (1959) method the same as those judged advanced or delayed by the Tanner, Whitehouse, Marshall, Healy, and Goldstein (1975) method? Both methods evaluate the maturity of the hand and wrist so that one can also inquire about relationships with other systems for assessing skeletal maturity, for example, the Roche, Wainer, and Thissen (1975) method of assessing skeletal maturity of the knee. Discrepancies also may be as large as 1 or more years between the skeletal ages of the knee and of the hand and wrist in individual children (Roche et al., 1975). Nevertheless, the hand and wrist complex is the most commonly used area for assessing skeletal maturity during growth. Finally, peak height velocity, another maturity indicator, requires longitudinal data, which present specific logistical problems in data collection and analysis (Malina, 1979a).

During adolescence maturity-associated variation in growth and body composition is most apparent. Also during this time, competitive sports are of primary importance to many youth. Hence, the interrelatedness of maturity and growth during adolescence is of importance. Indices of sexual and skeletal maturity and of growth are positively related during adolescence (Tanner, 1962; Malina, 1978). Most boys, for example, are in genital stage 4 and pubic hair stages 3 and 4 at peak height velocity, while most girls are in breast and pubic hair stage 3 at peak height velocity, and in breast and pubic hair stage 4 at menarche (Marshall, 1978). Menarche almost invariably occurs after peak height velocity, indicating the generally late occurrence of this developmental landmark within the matrix of adolescent events.

Results of principal components analyses of indices of somatic, sexual, and skeletal maturity in boys and girls of the Wrocław Growth Study (Bielicki, 1975; Bielicki et al., 1984) are summarized in Table 1. Both analyses resulted in two principal components. The first, which accounted for 77% and 68% of the total sample variance in boys and girls respectively, had high loadings from most indices. It thus suggests a general maturity factor, which discriminates among individuals who are early-, average-, or late-maturing. The second principal component accounted for 12% and 7.5% of the total sample variance in boys and girls, respectively. Among boys, this component, which had high-positive loadings on skeletal maturity at 11 and 12 years, and lower loadings at 13, 14, and 15 years, appears to be a rather specific factor that refers to the rate of skeletal maturation during preadolescence. The second component in girls is somewhat similar in that it had moderate loadings

Table 1. Principal components of ages at attaining certain developmental landmarks in the Wrocław Growth Study[a]

| Age at attaining: | Principal components | | | |
| | Boys | | Girls | |
	I	II	I	II
Peak height velocity	.95	−.19	.91	.09
Peak weight velocity	.89	−.23	.81	.13
Peak leg length velocity	.93	−.12	.89	.02
Peak trunk length velocity	.93	−.21	.89	.11
Skeletal maturity 11/10 years[b]	.51	.82	−.71	.53
Skeletal maturity 12/11 years	.62	.76	−.82	.48
Skeletal maturity 13/12 years	.82	.50	−.87	.41
Skeletal maturity 14/13 years	.92	.24	−.89	.32
Skeletal maturity 15/14 years	.95	−.02	−.88	.12
80% of adult stature	.77	.27		
Genital II/Breast II	.89	−.22	.85	.10
Genital IV/Breast IV	.91	−.17	.84	.21
Pubic hair II	.89	−.23	.85	.23
Pubic hair IV	.90	−.20	.84	.26
Menarche			.85	.17
Eigenvalues	13.04	2.00	12.89	1.41
Percent of variance explained	77%	12%	68%	7.5%

[a]Data compiled from Bielicki (1975) and Bielicki et al. (1984), based on 177 boys and 234 girls.

[b]For boys, skeletal maturity is the age at which the boys reached the median skeletal maturity score (Tanner-Whitehouse-Healy) for each chronological age; for girls, it is the median skeletal maturity score at each chronological age. Hence, a high score for girls indicates earlier maturation and is related with lower ages of attaining other indicators, thus negative loadings.

for skeletal maturity scores that systematically decreased from 10 through 14 years. The second principal component, however, differed between the sexes, particularly in the magnitude and sign of the loadings for indices of linear growth and secondary sex characteristic development.

Although there is some variation in the results of the two analyses of inter-relationships among indicators of somatic, sexual, and skeletal maturity, they have two implications. First, indices of linear growth, sexual maturation, and skeletal maturation are sufficiently interrelated to indicate a general maturity factor during adolescence. Second, in contrast to the notion of a general maturity factor, the evidence also suggests a certain degree of variation within and between the somatic, sexual, and skeletal indicators of growth and maturity. Among boys, the variation centered on the early stages of skeletal maturity as distinct from the later stages and other indicators. Among girls, the variation centered on the early stages of skeletal maturity and the later stages of secondary sex characteristic development. Hence, although there is a general maturity factor underlying the tempo of growth and maturation during adolescence in both sexes, there is variation so that no single system, that is, somatic growth, skeletal maturation, or sexual maturation, provides a complete description of the tempo of growth and maturation of an individual boy or girl during adolescence.

The apparent "disharmony" among indicators probably reflects differences in the nature of hormonal control between the prepubertal and pubertal phases of skeletal maturation. This is suggested by the correlations between the timing of take-off of the adolescent spurt, ages of peak velocities, and skeletal maturity at successive ages (see Table 2). The correlations increase regularly with increasing levels of skeletal maturity. As a result, age at take-off is more closely correlated with the age at reaching the skeletal maturity scores for 15 years of age in boys than it is with the age at reaching the skeletal maturity score for 12 years of age. The same is true in girls. The age at take-off is more highly related to the skeletal maturity score at 14 years of age than it is to the skeletal maturity score at 10 years. In other words, the age at take-off is more related to an event which occurs, on the average, more than 3 or 4 years later than with an event that, on the average, almost coincides with the take-off (Bielicki, 1975; Bielicki et al., 1984). The earlier phases of skeletal maturation are principally dependent upon the stimulation of growth hormone, while the later phases, which include epiphyseal capping and fusion, are chiefly under the influence of steroid hormones among others (Tanner et al., 1975). The later phases of skeletal maturation are thus under the control of the same factors which trigger the adolescent growth spurt and sexual maturation.

The preceding is consistent with the notion of central control of the tempo of maturation during adolescence. It is also consistent with the neuroendocrine hypothesis for the regulation of the adolescent growth spurt and sexual maturation (Grumbach, 1978, 1980; Kulin, 1974; Sizonenko, 1981). The hypothesis postulates a genetically programmed central nervous system (CNS) mechanism, including a "highly sensitive hypothalamic-pituitary-gonadal negative feedback mechanism" and " 'intrinsic' CNS inhibitory influences, independent of gonadal steroid feedback, on gonadotropin secretion . . ." (Grumbach, 1980, p. 254). At puberty, the hypothalamic-pituitary-gonadal system is activated,

Table 2. Correlations between skeletal maturity at successive ages and the timing of take-off and peak velocities in Wrocław boys and girls[a]

SM[b]	Boys ages at				SM[b]	Girls ages at			
	Take-off	PHV	PLLV	PTV		Take-off	PHV	PLLV	PTV
11	.19	.26	.35	.23	10	.46	.58	.61	.56
12	.33	.42	.49	.39	11	.55	.69	.69	.67
13	.57	.68	.73	.64	12	.58	.73	.71	.73
14	.71	.81	.83	.78	13	.62	.76	.75	.76
15	.78	.89	.89	.87	14	.68	.79	.78	.77

[a]Data compiled from Bielicki (1975) and Bielicki et al. (1984).
Take-off = initiation of adolescent spurt; PHV = peak height velocity; PLLV = peak leg length velocity; PTV = peak trunk velocity.
[b]For boys, skeletal maturity (SM) is the age at which the boys reached the median skeletal maturity score for each chronological age; for girls, it is the median skeletal maturity score at each chronological age. Hence, for girls, the signs of the correlations were reversed.

while the CNS inhibitory influence decreases. The process, however, is gradual. The neuroendocrine changes begin, on the average, at about 9 or 10 years of age in girls and 10 or 11 years in boys and continue through the growth spurt and sexual maturation. Although a CNS mechanism apparently triggers these changes, there is most likely variation in target organ responsiveness to the hormonal changes. Variation in hormone receptor concentration may, for example, underlie some of the variation among indicators of somatic, sexual, and skeletal maturity during adolescence.

Training and Maturation

The central control of the adolescent growth spurt and sexual maturation is genetically determined: "When socioeconomic and environmental factors are optimal for growth and development, the age of onset of puberty in normal children is determined principally by genetic factors" (Grumbach, 1980, p. 250). Indeed, evidence from twin studies indicates high heritabilities for secondary sex characteristic development, menarche, peak height velocity, age at peak height velocity, and skeletal maturation (Fischbein, 1977a, 1977b; Kimura, 1981, 1983; Skład, 1973, 1977). The pattern of plasma hormonal levels also shows a significant genotypic component in monozygotic pubescent twins (Parker, Judd, Rossman, & Yen, 1975). However, the timing of the growth spurt and puberty can be influenced by environmental factors, for example, chronic undernutrition and anorexia nervosa. Evidence from postpubertal anorexia nervosa patients indicates a reversion of gonadotropin and estradiol secretion to the prepubertal state (Grumbach, 1980). A logical question is whether regular training can prolong the prepubertal state and hence delay sexual maturation and related somatic and skeletal maturation.

Delayed Menarche

The available data focus primarily on delayed menarche in athletes and by inference the effects of training (Malina, 1983a). Other maturity indicators are generally not used, while training-maturity associations in males are not considered even though the hormonal changes are principally the same.

The suggested mechanism for the association between intensive physical training and delayed menarche in athletes is hormonal. The data dealing with the relationship between training and delayed menarche, however, are associational, limited, quite speculative, and do not control for other factors which are known to influence menarche (Malina, 1983a). For example, in a small sample ($n = 18$) of university level swimmers and runners, Frisch and colleagues (1981) observed a correlation of $+.53$ between age at menarche and years of training before menarche. Although this moderate correlation accounts for only about 28% of the sample variance, Frisch et al. (1981, p. 1562) concluded that "intense physical activity (before menarche) does in fact delay menarche." More recently, Stager, Robertshaw, and Miescher (1984) reported a low correlation of $+.32$ (which accounts for about 10% of the sample variance) between years of training before menarche and the age of menarche in a large sample of swimmers and active nonswimmers. In contrast, our data for a small sample ($n = 18$) of Olympic volleyball players indicate no correlation ($r = -.05$) between age at menarche and duration of training before menarche (Malina, 1983a).

Obviously, correlation does not imply a cause-effect relationship. It could well be that lateness in menarche may motivate girls to take up sport training rather than the training causing the lateness (Malina, 1983a). Further, the association may be an artifact. The older a girl is at menarche, the more likely she would have begun her training prior to menarche. Conversely, the younger a girl is at menarche, the more likely she would have begun training after menarche (Stager et al., 1984).

It is suggested that intensive training and perhaps associated energy drain influence circulating levels of gonadotrophic and ovarian hormones and, in turn, menarche. Exercise is a highly effective means of stressing the hypothalamic-pituitary-gonadal axis. Most of the data, however, are derived from studies of women (both athletes and nonathletes) who have attained menarche (Malina, 1983a). An exception is the observation of Warren (1980), who noted extremely low gonadotropin secretion associated with only "mild" growth stunting in premenarcheal ballet dancers. Otherwise, the evidence indicates short-term exercise-related increases in serum levels of almost all gonadotrophic and sex steroid hormones (Cumming, Brunsting, & Greenberg, 1981; Shangold, 1984; Terjung, 1979). The many factors, in addition to exercise, capable of influencing hormonal levels must also be considered, for example, diurnal variation, state of feeding or fasting, emotional states, and so on (Shangold, 1984). In addition, many hormones are secreted in a pulsatile fashion so that single serum samples may not reflect the overall pattern.

Hormonal Responses

The acute hormonal responses to exercise are apparently essential to meet the stress that intense exercise imposes upon the organism. More information is needed on changes, if any, in basal levels of hormones with training. Strength training, for example, does not apparently change resting levels of plasma testosterone in adult men and women (Hetrick & Wilmore, 1979).

In children, the increased production of gonadotrophic hormones with the onset of puberty occurs primarily during sleep. The periodic release of luteinizing hormone (LH) is correlated with the number of sleep cycles in late prepubertal or early pubertal boys and precedes the initial overt signs of puberty. Apparently, sleep has

> A functional and not merely an incidental role in normal development. Since growth hormone, prolactin, and LH during puberty are secreted primarily during sleep, it is possible that sleep is an important mechanism for synchronizing the metabolic and tissue changes that occur during puberty. (Weitzman, 1980, p. 92)

The effects of training and competition on sleep in young athletes thus merit closer scrutiny. The effects of daytime exercise on sleep are presumably beneficial, that is, longer sleep, more slow-wave sleep, and more growth hormone production during sleep (Oswald, 1980). On the other hand, stress is associated with both increased and decreased LH concentrations in the waking state (Shangold, 1984). The effects of pre- and postcompetition anxiety on sleep and, of course, on hormone production might thus be a seemingly contradictory factor.

A role for beta-endorphins in amenorrhea of runners (McArthur et al., 1980) and, in turn, delayed menarche in athletes (McArthur, 1983) has been postulated (see also Shangold, 1984). The effect of *naloxone* (an opiate receptor antagonist) under conditions of exercise in adults, for example, results in a marked increase in LH. However, the response of normal prepubertal boys and girls to naloxone under basal conditions is different from that of adults (Fraioli et al., 1984). Naloxone apparently does not have an effect on LH secretion in children. A study of the effects of naloxone under exercise conditions in children thus seems to be warranted; however, ethical considerations would dictate otherwise.

It is somewhat puzzling why one would expect training to delay the maturation of girls and not boys, even though the underlying processes are quite similar. This is clearly evident in the literature. With the exception of studies of growth hormone responses to exercise, data on other hormonal responses of boys to exercise are generally limited. One of the few studies (Fahey, del Valle-Zuris, Oehlsen, Trieb, & Seymour, 1979) did not observe any differences in postmaximal exercise concentrations of serum testosterone and growth hormone among boys grouped into different stages of puberty. This would suggest that a critical pubertal state with enhanced hormonal responsiveness to exercise does not exist, at least in boys.

Most of the literature on hormonal responses to exercise is based upon observations of serum or plasma levels. However, the simple presence of a hormone does not necessarily imply that it is physiologically or biochemically active. There also is variation in tissue responsiveness, and in developing children, the tissue probably must be sufficiently mature in order to respond. In other words, for a hormone to have its effect, the tissue, or more specifically, the cells, must be capable of responding.

As noted earlier, a moderately high correlation exists among the development of secondary sex characteristics, peak height velocity, and skeletal maturity in both sexes. In girls, of course, the age at menarche is also highly related to other indices of maturity, and there is a reduced variance in skeletal age at menarche (Simmons, 1944; Marshall, 1974). Hence, if the hormonal responses to regular training are viewed as important influences on sexual maturation, one might expect them to influence skeletal maturation, especially during puberty, because epiphyseal capping and fusion are influenced by gonadal hormones, among others. The data, however, are limited.

Delayed Skeletal Maturity

Delayed skeletal maturity is commonly observed in female gymnasts and ballet dancers. And in the case of the latter, it has been observed in association with extremely low levels of gonadotrophins (Warren, 1980). Changes in skeletal maturity of the hand and wrist in association with regular training, on the other hand, show negligible training effects in both sexes.

Novotny (1981), for example, compared the progress of skeletal maturation in elite Czechoslovak athletes in several sports over periods of 3 to 4 years between 12 and 17 years of age. Athletes were grouped as having accelerated, average, or delayed skeletal ages at the start and at the end of the study. Changes in categories were subsequently reported (see Table 3). There was little variation in category changes among sports after the period of regular training. Only 19 (21%) of the 89 girls changed categories, while 70 (79%) remained in the same skeletal maturity category. Given the error in assessing skeletal ages and normal variation in the timing of the adolescent spurt and sexual maturation, the changes are quite small. Of the small number who changed categories, 11 shifted from advanced to normal or from normal to delayed, while eight shifted from delayed to normal or from normal to advanced. In addition, at the beginning and end of the study, mean chronological and skeletal ages of the young athletes did not differ significantly. Hence, gains in chronological and skeletal age were also very similar over the period of training and competition.

Among boys, Černy (1969) compared the skeletal maturity of boys undergoing different training programs over a 4-year period from 11 to 15 years, a period during which most boys experience their growth spurt and sexual maturation. Three levels of training were compared: regularly trained (i.e., intensive, 6 hours per week); trained but not on a regular basis (i.e., went to sport schools, about 4 hours of organized exercise per week); and untrained (i.e., about 2-1/2 hours per week inlcuding school physical education). A fourth group, including some dropouts from the training programs, was irregularly active. It should be noted, however, that the regularly trained boys were not nearly as active as competitive age group swimmers or runners. Both chronological age and skeletal age made, on the average, similar progress over the 4 years in the different activity groups. Similar results were reported by Kotulán, Řezničková, and Placheta (1980) on young males training regularly for cycling, rowing, and ice hockey from 12 to 15 years of age (see Table 4). Over the 3-year period, the gains in skeletal maturity varied, on the average, between 2.6 and 3.3 years in the athletes, and did not differ from the control subjects and from youngsters who trained irregularly during the project.

The preceding would suggest, therefore, that the process of skeletal maturation as reflected in the hand and wrist is not affected by regular physical training for sport in adolescent boys and girls. It may also be inferred from these observations that skeletal maturity is stable over time, that is, those advanced at one age are most likely advanced at later ages, and vice versa. Note, however, that as skeletal maturity of the hand and wrist is attained, approximately 16 years in girls and 18 years in boys, the differences between those advanced and delayed in maturity are reduced and eventually eliminated.

Implications

The apparent lack of significant training effects on indices of maturation seems to emphasize the role of constitutional factors in successful sports participation at young ages. For example, among two small samples of boys who had no sports training prior to 11 years of age, those who started to train in track

Table 3. Skeletal maturity in elite young female athletes: Change over time[a]

Sport	n	Mean chronological ages		Gain in CA[b]	Gain in SA[b]	Change in skeletal age category None n	(+) n	(-) n
Gymnastics	24	12.3	16.5	4.2	4.1	17	3	4
Figure skating	16	12.6	16.1	3.5	4.0	10	3	3
Tennis	14	14.3	17.1	2.8	2.5	12	0	2
Volleyball	12	14.2	17.0	2.8	2.5	10	0	2
Football (soccer)	23	13.5	16.8	3.3	3.3	21	2	0
Total	89					70	8	11

aData compiled from Novotny (1981).
bDifferences between mean chronological age (CA) and skeletal age (SA), respectively, at the beginning and the end of the study.

Table 4. Skeletal maturity in young male athletes: Gain in years[a]

Years	Cycling (n = 6) CA	SA	Rowing (n = 11) CA	SA	Ice hockey (n = 16) CA	SA	Various (n = 17) CA	SA	Dropouts (n = 19) CA	SA	Controls (n = 34) CA	SA
1972-1973	.95	.91	.92	1.02	1.02	1.01	.96	1.03	.94	1.11	1.01	1.07
1973-1974	1.02	.79	1.05	1.34	.99	.86	1.04	.99	1.01	1.02	.98	.93
1974-1975	.96	.88	.97	.97	1.00	1.10	.98	1.27	1.00	1.15	.99	1.14
1972-1975	2.93	2.58	2.94	3.33	3.01	2.97	2.98	3.29	2.95	3.28	2.98	3.22

aData compiled from Kotulán et al. (1980). Gains are the differences between mean CA and mean SA, respectively, in adjacent years and at the beginning and the end of the study.

and basketball and who persisted in training between 11 and 18 years of age were consistently advanced in skeletal maturity. In contrast, those who chose not to train were consistently delayed in skeletal maturity (Ulbrich, 1971). Likewise, the early survey of Baker (1940) would seem to imply constitutional factors. In a survey of factors influencing participation in physical education activities by young adult women, late menarche was significant. Those young women who attained menarche after 15 years of age participated more than those attaining menarche earlier.

Constitutional factors imply size, physique, body composition, maturity, and performance variation among individuals, all of which have a significant genotypic contribution (Bouchard & Lortie, 1984; Bouchard & Malina, 1983). Training is a significant factor influencing body composition and performance, but not size (stature), physique, or maturation (Malina, 1979b, 1983b).

Growth and maturity characteristics of young athletes have been previously reviewed (Malina, 1982, 1983a; Malina, Meleski, & Shoup, 1982). For males, the data suggest that early maturation with its concomitant size and strength advantages constitutes an asset positively associated with success in several sports in early adolescence (11-14 years), for example, American football, baseball, and swimming. Ice hockey, on the other hand, is an apparent exception. Boys average or delayed in skeletal maturity are more commonly successful at young ages. Data for European football (soccer) are not extensive. Some evidence at the local club level indicates no clear maturity relationships (Vrijens & Van Cauter, 1983). Evidence on successful young long-distance runners is presently being gathered (Vogel, 1983).

As adolescence approaches its termination, the maturity status of the young males is of less significance. The deceleration in growth and maturation of early-maturing boys and the catch-up of late-maturing boys reduces the size differences so apparent in early adolescence, although physique differences persist. Hence, a question that merits study is, Who are the successful athletes in early and late adolescence? The sample comprising young male athletes at the pre- and early-adolescent ages may be different from that in late adolescence. For example, in small samples of all-star ice hockey players and chronological age-matched controls, Hamilton and Andrew (1976) noted that the prepubertal all-star hockey players did not differ significantly in size from the age-matched prepubertal controls. The hockey players were, in fact, slightly shorter and lighter on the average. However, at a postpubertal age (16 years), all-star hockey players were significantly taller and heavier and had more weight for height than the chronological age-matched control subjects. These observations, though limited to small samples, suggest that the population of successful ice hockey players in late adolescence may be different from that during pre- or early-adolescence.

Among girls, the data indicate delayed biological maturity in athletes compared to nonathletes with the possible exception of swimmers. Gymnasts, figure skaters, and ballet dancers are most delayed in maturity. However, most sports accessible to young girls are individual sports with a smaller number of participants and with perhaps more rigorous selection criteria at the elite competitive levels in contrast to team sports which are ordinarily more available for boys. Casual observations of participants in youth sports at the local level, however, would seem to indicate trends similar to those evident in boys. In Little League softball for girls, for example, the more successful players are more often than not larger in size and advanced in maturity status. The size advantage is especially evident in batting and throwing. Hence, as in the case for boys, there is a need to study the population of young athletes in different sports, at different competitive levels, and at different ages. Further, with Title IX legislation in the United States, many elite female athletes now have the opportunity to compete at the high school and university levels. Hence, there also is a need to study female participants in a greater variety of sports, especially as the opportunity to participate improves.

The preceding information thus implies the need to consider other factors in successful sport performance. Successful young athletes are sometimes labeled as gifted, and it is of interest to inquire whether they share characteristics in common with youngsters successful in other competitions, such as music, the arts, and mathematics. Bloom (1982) did a detailed retrospective study of individuals who attained "world-class" status at a relatively early age (as early as 17) and as late as 35 years. The samples included Olympic swimmers, concert pianists, and research mathematicians. Successful individuals in the diverse areas of sport, music, and mathematics showed three characteristics in common: willingness to do great amounts of work aimed at high goals, great competitiveness, and ability to learn rapidly. The successful swimmers demonstrated two additional characteristics: ease in water and a special "feel" for the water. The physical characteristics of swimmers are also postulated as important, primarily in providing early competitive advantages and in securing expert coaching:

> Natural physical characteristics that give an individual some initial advantage over his or her age mates are likely to function to motivate the individual to enter and compete in a sport. They also help him or her secure the teaching and training needed to convert an individual with small initial advantages into a world-class athlete. (Bloom, 1982, p. 515)

However, these native "gifts" are probably not as significant at later ages when international competition occurs. At such levels, athletes have much in common, so that other factors, many quite subtle, are important in obtaining the so-called competitive edge.

Summary

Many factors are undoubtedly related to successful athletic performance. Variation in the rate of biological maturation may be of significance in youth sport, perhaps in providing early competitive advantages in some sports. However, with time during adolescence, maturity-associated variation in most characteristics is reduced so that the sample of athletes in some sports in late adolescence may be different from that in pre- or early-adolescence.

Individual variation in biological maturation and associated changes in size, physique, body composition, and performance are the backdrop against which youth evaluate and interpret their own growth and maturation. The adolescent growth spurt and sexual maturation do not occur in a social vacuum. Physical activity and sport are an important part of the youngster's world, especially during adolescence, and success in athletics is a significant form of positive social reinforcement in the culture of youth.

It is difficult to implicate the stress of training and competition as a critical influence on biological maturation. Hypotheses on the effect of training on young female athletes need to consider other factors that are known to influence menarche, which is a late maturational event, as well as other factors that are known to influence circulating gonadotrophic and gonadal hormone levels. More importantly, the mere presence of a hormone in circulation does not guarantee that it is physiologically or biochemically active. The tissues must be capable of responding to the hormone.

In contrast, the effect of training and competition on young male athletes often goes unnoticed, although the biological processes that underlie sexual maturation of males are quite similar to those in females. Successful young male athletes in some sports are advanced in biological maturity status, while in other sports they are delayed or average. Yet, the stress of training is not generally considered a significant factor affecting the maturation of young males.

References

Baker, M.C. (1940). Factors which may influence the participation in physical education of girls and women 15 to 25 years of age. *Research Quarterly, 11*, 126-131.

Bielicki, T. (1975). Interrelationships between various measures of maturation rate in girls during adolescence. *Studies in Physical Anthropology, 1*, 51-64.

Bielicki, T., Koniarek, J., & Malina, R.M. (1984). Interrelationships among certain measures of growth and maturation rate in boys during adolescence. *Annals of Human Biology, 11*, 201-210.

Bloom, B.S. (1982). The role of gifts and markers in the development of talent. *Exceptional Children, 48*, 510-522.

Bouchard, C., & Lortie, G. (1984). Heredity and endurance performance. *Sports Medicine, 1*, 38-64.

Bouchard, C., & Malina, R.M. (1983). Genetics of physiological fitness and motor performance. *Exercise and Sport Sciences Reviews, 11*, 306-339.

Černy, L. (1969). The results of an evaluation of skeletal age of boys 11-15 years old with different regimes of physical activity. In V. Novotny (Ed.), *Physical fitness assessment* (pp. 56-59). Prague: Charles University.

Cumming, D.C., Brunsting, L., Greenberg, L., et al. (1981). Patterns of endocrine response to exercise in normal women. Cited by R.W. Rebar and D.C. Cumming (1981) in the *Journal of the American Medical Association, 246*, 1590. (From the Sixty-third Annual Meeting, June 17-19, Cincinnati, OH, Abstract No. 465, p. 199.)

Demirjian, A. (1978). Dentition. In F. Falkner & J.M. Tanner (Eds.), *Human growth. Volume 2. Postnatal growth* (pp. 413-444). New York: Plenum.

Fahey, T.D., del Valle-Zuris, A., Oehlsen, G., Trieb, M., & Seymour, J. (1979). Pubertal stage differences in hormonal and hematological responses to maximal exercise in males. *Journal of Applied Physiology, 46*, 823-827.

Fischbein, S. (1977a). Onset of puberty in MZ and DZ twins. *Acta Geneticae, Medicae et Gemellologiae, 26*, 151-157.

Fischbein, S. (1977b). Intra-pair similarity in physical growth of monozygotic and of dizygotic twins during puberty. *Annals of Human Biology, 4*, 417-430.

Fraioli, F., Cappa, M., Fabbri, A., Gnessi, L., Moretti, C., Borrelli, P., & Isidori, A. (1984). Lack of endogenous opioid inhibitory tone on LH secretion in early puberty. *Clinical Endocrinology, 20*, 299-305.

Frisch, R.E., Gotz-Welbergen, A.V., McArthur, J.W., Albright, T., Witschi, J., Bullen, B., Birnholz, J., Reed, R.B., & Hermann, H. (1981). Delayed menarche and amenorrhea of college athletes in relation to age of onset of training. *Journal of the American Medical Association, 246*, 1559-1563.

Greulich, W.W., & Pyle, S.I. (1959). *Radiographic atlas of skeletal development of hand and wrist* (2nd ed.). Stanford: Stanford University Press.

Grumbach, M.M. (1978). The central nervous system and the onset of puberty. In F. Falkner & J.M. Tanner (Eds.), *Human growth. Volume 2. Postnatal growth* (pp. 215-238). New York: Plenum.

Grumbach, M.M. (1980). The neuroendocrinology of puberty. In D.T. Krieger & J.C. Hughes (Eds.), *Neuroendocrinology* (pp. 249-258). Sunderland, MA: Sinauer Associates.

Hamilton, P., & Andrew, G.M. (1976). Influence of growth and athletic training on heart and lung functions. *European Journal of Applied Physiology, 36*, 27-38.

Hetrick, G.A., & Wilmore, J.H. (1979). Androgen levels and muscle hypertrophy during an eight week training program for men/women. *Medicine and Science in Sports, 11*, 102.

Kimura, K. (1981). Skeletal maturity in twins. *Journal of the Anthropological Society of Nippon, 89*, 457-477.

Kimura, K. (1983). Skeletal maturity and bone growth in twins. *American Journal of Physical Anthropology, 60*, 491-497.

Kotulán, J., Řezničková, M., & Placheta, Z. (1980). Exercise and growth. In Z. Placheta (Ed.), *Youth and physical activity* (pp. 61-117). Brno: J.E. Purkyne University Medical Faculty.

Kulin, H.E. (1974). The physiology of adolescence in man. *Human Biology, 46*, 133-144.

Malina, R.M. (1978). Adolescent growth and maturation: Selected aspects of current research. *Yearbook of Physical Anthropology, 21*, 63-94.

Malina, R.M. (1979a). Longitudinal growth studies: Approaches, problems and results. In D. Mood (Ed.), *The measurement of change in physical education* (pp. 125-157). Boulder: University of Colorado.

Malina, R.M. (1979b). The effects of exercise on specific tissues, dimensions, and functions during growth. *Studies in Physical Anthropology, 5*, 21-52.

Malina, R.M. (1982). Physical growth and maturity characteristics of young athletes. In R.A. Magill, M.J. Ash, & F.L. Smoll (Eds.), *Children and sport* (2nd ed.) (pp. 73-96). Champaign, IL: Human Kinetics.

Malina, R.M. (1983a). Menarche in athletes: A synthesis and hypothesis. *Annals of Human Biology, 10*, 1-24.

Malina, R.M. (1983b). Human growth, maturation, and regular physical activity. *Acta Medica Auxologica, 15*, 5-27.

Malina, R.M. (1984). Physical growth and maturation. In J.R. Thomas (Ed.), *Motor development during childhood and adolescence* (pp. 2-26). Minneapolis: Burgess.

Malina, R.M., Meleski, B.W., & Shoup, R.F. (1982). Anthropometric, body composition, and maturity characteristics of selected school-age athletes. *Pediatric Clinics of North America, 29*, 1305-1323.

Marshall, W.A. (1974). Interrelationships of skeletal maturation, sexual development and somatic growth in man. *Annals of Human Biology, 1*, 29-40.

Marshall, W.A. (1978). Puberty. In F. Falkner & J.M. Tanner (Eds.), *Human growth. Volume 2. Postnatal growth* (pp. 141-181). New York: Plenum.

McArthur, J.W. (1983). *Endorphins and exercise in females*. Paper presented at the 1983 Annual Meeting of the American College of Sports Medicine, Montreal.

McArthur, J.W., Bullen, B.A., Beitins, I.Z., Pagano, M., Badger, T.M., & Klibanski, A. (1980). Hypothalamic amenorrhea in runners of normal body composition. *Endocrine Research Communications, 7*, 13-25.

Novotny, V. (1981). Veränderungen des Knochenalters im Verlauf einer mehrjährigen sportlichen Belastung. *Medizin und Sport, 21*, 44-47.

Oswald, J. (1980). Sleep as a restorative process: Human clues. In P.S. McConnell, G.J. Boer, H.J. Romijn, N.E. van de Poll, & M.A. Corner (Eds.), *Adaptive capabilities of the nervous system* (pp. 279-288). New York: Elsevier.

Parker, D.C., Judd, H.L., Rossman, L.G., & Yen, S.S.C. (1975). Pubertal sleep-wake patterns of episodic LH, FSH and testosterone release in twin boys. *Journal of Clinical Endocrinology and Metabolism, 40*, 1099-1109.

Roche, A.F., Wainer, H., & Thissen, D. (1975). *Skeletal maturity: The knee joint as a biological indicator*. New York: Plenum.

Shangold, M.M. (1984). Exercise and the adult female: Hormonal and endocrine effects. *Exercise and Sport Sciences Reviews, 12*, 53-79.

Simmons, K. (1944). The Brush Foundation study of child growth and development. II. Physical growth and development. *Monographs of the Society for Research in Child Development, 9*(1).

Sizonenko, P.C. (1981). Regulation of puberty and pubertal growth. In M. Ritzén, A. Aperia, K. Hall, A. Larsson, A. Zetterberg, & R. Zetterström (Eds.), *The biology of normal human growth* (pp. 297-308). New York: Raven Press.

Skład, M. (1973). Genetische Grundlagen einiger mit dem Heranreifen verbundener Erscheinungen (nach Zwillingsuntersuchungen). *Zeitschrift für Morphologie und Anthropologie, 65*, 192-211.

Skład, M. (1977). The rate of growth and maturing of twins. *Acta Geneticae, Medicae et Gemellologiae, 26*, 221-237.

Stager, J.M., Robertshaw, D., & Meischer, E. (1984). Delayed menarche in swimmers in relation to age at onset of training and athletic performance. *Medicine and Science in Sports and Exercise, 16*, 550-555.

Tanner, J.M. (1962). *Growth at adolescence* (2nd ed.). Oxford: Blackwell Scientific Publications.

Tanner, J.M., Whitehouse, R.H., Marshall, W.A., Healy, M.J.R., & Goldstein, H. (1975). *Assessment of skeletal maturity and prediction of adult height*. New York: Academic Press.

Terjung, R. (1979). Endocrine response to exercise. *Exercise and Sport Sciences Reviews, 7*, 153-180.

Ulbrich, J. (1971). Individual variants of physical fitness in boys from the age of 11 up to maturity and their selection for sports activities. *Medicina dello Sport, 24*, 118-136.

Vogel, P.G. (1983). Youth Sports Institute studies the effects of long distance running on children. *Spotlight on Youth Sports, 6*(1), 1, 3.

Vrijens, J., & Van Cauter, C. (1983). *Physical performance capacity and specific skills in young soccer players of ages 10 to 13 years*. Paper presented at Pediatric Work Physiology XI, Papendal, The Netherlands.

Warren, M.P. (1980). The effects of exercise on pubertal progression and reproductive function in girls. *Journal of Clinical Endocrinology and Metabolism, 51*, 1150-1157.

Weitzman, E.D. (1980). Biological rhythms and hormone secretion patterns. In D.T. Krieger & J.C. Hughes (Eds.), *Neuroendocrinology* (pp. 85-92). Sunderland, MA: Sinauer Associates.

3

The Relationship of Chronological Age, Body Build, and Sexual Maturation to Handgrip Strength in Schoolboys Ages 10 Through 17 Years

Flávia C. Bastos and Raymond V. Hegg
UNIVERSIDADE DE SÃO PAULO
SÃO PAULO, BRAZIL

During adolescence, the capacity for and organic adaptation to physical activities depends closely on the occurrence of qualitative changes. These changes are a result of neurohormonal alterations (Tanner, 1962). Specifically, the spurt of strength development is often considered as an indicator of maturity because muscular tissue increases first in mass and later in strength, which suggests a qualitative change in the muscular tissue, or maturation resulting from development (Malina, 1978). Malina and Mueller (1981) stressed the relevance of body-build relationships, stated in weight and height, and strength scores. Somatic changes lead to important alterations in the functional pattern of the organism and allow for an increase in physical capacity.

A recent approach relates maturational level to the development of muscular strength in adolescents. Studies using maturational variables have confirmed the important relationship between the discharge of large amounts of testosterone after the onset of puberty and the rapid growth of muscular mass, resulting in a significant increase in strength in males (Beunen, 1973; Hebbelinck & Borms, 1975; Nakagawa, Takahashi, Tomabechi, & Nakamura, 1958). This chapter describes a study which investigated the influence of chronological age, some body-build variables, and sexual maturation on the handgrip strength of Brazilian adolescents, and identified the most significant variable during this period.

Method

The subjects were 363 upper middle-class schoolboys, aged 10 to 17 years, attending a private school in São Paulo, Brazil. The variables used were weight, height, body surface area, right-handgrip strength, and sexual maturation as determined by pubic hair. The data were collected over a 2-month period. Chronological age was recorded in months, and the age groups were based on birth date and test date. Weight and height were used to determine the body surface area in accordance with the formula developed by Du Bois and Du Bois (1916). Pubic hair development was determined in six stages in accordance with the standards published by Tanner (1962). Handgrip strength was measured by a TKK Grip Dynamometer adjustable to hand size. Two trials for handgrip strength were administered and the highest score was recorded. Stepwise multiple regression analysis was used for the prediction of handgrip strength of the entire group and for each age group separately, with the level of significance set at .05.

Results and Discussion

The results of the multiple regression analysis for the entire group and the standardized multiple regression coefficients are shown in Table 1. The first step included only the pubic hair stage ($R = .87$), which indicates that this variable was a significant estimator of handgrip strength. This variable accounted for 76.33% ($R^2 \times 100$) of the handgrip variation. The findings from Nakagawa, Takahashi, Tomabechi, and Nakamura (1958) regarding develop-

Table 1. Regression coefficients for prediction of handgrip strength in boys[a] aged 10 to 17 years

| Independent variables | Step 1 | Standardized coefficients | | |
		Step 2	Step 3	Step 4
Pubic hair stage	0.87*	0.51*	0.47*	0.42*
Height		0.44*	0.28*	0.20*
Surface area			0.21*	0.24*
Chronological age				0.11*
R	0.87*	0.90*	0.91*	0.92*
$R^2 \times 100$	76.33	82.58	83.11	83.45

Note. *$p < .01$.
[a]$N = 363$.

mental stages of axillary hair and handgrip strength and from Hebbelinck and Borms (1975) on the development of pubic hair and handgrip strength in 12-year-old boys showed that groups of early-maturing boys demonstrated greater handgrip strength than those maturing later. These results confirm the close relationship between sexual maturation of adolescents and their handgrip strength. This relationship was also suggested by Tanner (1962) and Rarick (1973) based on the correspondence existing between the level of individual maturation and the increase in muscular mass, mainly in the upper extremities, which is due to considerable increases in the level of testosterone circulating after the onset of puberty.

Chronological age presented the lowest regression coefficient among the variables studied, evidencing its slight influence on the handgrip strength of the group as a whole. This result confirms the findings from Beunen (1973), Rarick (1973), and Teeple (1973) of a negligible relationship with handgrip strength.

Among the body size variables, height was found to be the best estimator of handgrip strength, followed by body surface area. The latter only slightly affected strength prediction, from 82.58 to 83.11% ($R^2 \times 100$) from the second to the third step. Weight did not satisfy the condition of significance established, although it had an indirect influence through the body surface area. In the study developed by Lamphiear and Montoye (1976), the regression analysis for the 10- to 15-year-old age group selected height, weight, biacromial diameter, triceps skinfold, and arm girth as the variables of greatest influence on the prediction of upper limb strength. Together, they accounted for approximately 81% of observed variation. Teeple (1973) did not find weight to be a significant variable in the prediction of handgrip strength in boys aged 6 to 12 years, although fat-free weight was the most important variable in her study.

The multiple regression analysis for each age group (see Table 2) showed that pubic hair appears as an influencing factor at the ages of 11, 12, 14, 15, and 16/17 years, but does not appear as an estimator at the age of 13 years.

Table 2. Regression coefficients for prediction of handgrip strength by age group

			Independent variables				
Age	n	$R^2 \times 100$	Weight	Height	Surface area	Pubic hair	Constant
10	57	51.54		0.72*			−32.72
11	62	46.98			0.32*	0.44*	3.17
12	55	58.11		0.54*		0.30**	−22.75
13	58	68.32		0.42**	0.43**		−39.78
14	46	51.48		0.41*		0.38*	−33.69
15	51	61.55	0.30*			0.57*	13.38
16/17	34	39.49		0.50*		0.31**	−51.95

Note. N = 363.
*p < .05.
**p < .01.

This factor entered the regression analysis at the age of 11 years, decreased from 11 to 12 years, disappeared at age 13, increased from 14 to 15 years (when it peaked), and decreased again from 15 to 16/17 years. Among the body-build variables, height appears as an influencing factor at 10, 12, 13, 14, and 16/17 years of age. The coefficient decreases in value at 10 through 13 years of age, remains steady at 14, and increases at 16/17 years of age. Body surface area and weight variables appear in isolated age groups: body surface area at 11 and 13 years, and weight at 15 years. Finally, some age trends were observed in the influence of body size variables and pubic hair on strength prediction. In general, there is a decreasing influence of height from 10 to 13 years of age, stabilizing at 14 years, and an increasing influence from 12 to 15 years of age for pubic hair.

The conclusions of the study were limited by the cross-sectional data. The findings, however, were in agreement with those of Tanner (1962) on the effects of testosterone on the strength of boys after the age of 13.5 years, and with those of Carron and Bailey (1974) and Malina (1978) on the occurrence of a strength growth spurt about 1 year after the maximum growth spurt in height and weight. That is to say, the increase in strength in individuals at 15 years of age depends more on sexual maturation than on body size.

Conclusions

Sexual maturation as determined by pubic hair development was the most significant factor in predicting handgrip strength in schoolboys and chronological age was the least influential factor. Also, during the period from 10 to 17 years of age, height has a decreasing influence from age 10 to 14 years, and sexual maturation has an increasing influence from age 12 to 15 years. Further studies are required to determine more precisely the influence of body size variables.

References

Beunen, G. (1973). Utilité de la détermination de la maturité osseuse lors de l'évaluation de l'aptitude physique de jeunes garcons. *Sport,* **4**(64), 220-231.

Carron, A.V., & Bailey, D.A. (1974). Strength development in boys from 10 through 16 years. *Monographs of the Society for Research in Child Development,* **39**(4) (Serial 157).

Du Bois, D., & Du Bois, E.F. (1916). A formula to estimate the approximate surface area if height and weight be known. *Archives of International Medicine,* **17**, 836-871.

Hebbelinck, M., & Borms, J. (1975). Puberty characteristics and physical fitness of primary school children aged 6 to 13 years. In S.R. Berenberg (Ed.), *Puberty, biologic and psychosocial components* (pp. 224-235). Leiden: Stenfert Kroese Publishers.

Lamphiear, D.E., & Montoye, H.J. (1976). Muscular strength and body size. *Human Biology,* **48**(1), 147-160.

Malina, R.M. (1978). Growth of muscle tissue and muscle mass. In F. Falkner & J.M. Tanner (Eds.), *Human growth* (pp. 273-294). New York: Plenum Press.

Malina, R.M., & Mueller, W.H. (1981). Genetic and environmental influences on the strength and motor performance of Philadelphia school children. *Human Biology, 53*(2), 163-179.

Nakagawa, I., Takahashi, T., Tomabechi, K., & Nakamura, I. (1958). Studies on growth and development during adolescence. *Bulletin of the Institute of Public Health, 7*(3), 161-174.

Rarick, G.L. (1973). Physical activity—human growth and development. New York: Academic Press.

Tanner, J.M. (1962). *Growth at adolescence*. Oxford: Blackwell Scientific Publications.

Teeple, J.B. (1973). *The influence of physical growth and maturation on the static force production of boys ages 6 through 12 years*. Unpublished doctoral dissertation, University of Illinois at Urbana-Champaign.

4

Relationships Between Skeletal Age and Physical Fitness Variables of 6-year-old Children

Marcel Hebbelinck, Jan Borms,
William Duquet, and Marleen Vanderwaeren
VRIJE UNIVERSITEIT BRUSSEL
BRUSSELS, BELGIUM

A great deal of information is available concerning the relationships between maturational development and physical fitness variables in adolescents (Malina, 1975), but analogous information in primary school children is rather scarce. Body size, proportions, and composition vary considerably with the maturity status of the developing individual in the adolescent period, but maturity-related variation in physique is also noted during childhood (Hebbelinck & Borms, 1978). Asmussen and Heeböl-Nielson (1955) reported that gains in muscular strength of boys aged 7 to 17 years were greater than the theoretical values based upon calculated indices of body structure. They suggested that this was due to some unidentified maturational influence.

Studies concerning the relationships between skeletal maturity and physical fitness variables in young school-age children have been carried out by Seils (1951) on first, second, and third grade (6 to 8.8 years) boys and girls, and by Rarick and Oyster (1964) on second grade (8.3 years) boys. Correlations between skeletal maturity and strength and gross motor performance test scores were positive, ranging from low to moderate. Partialing out the effects of chronological age, height, and weight, Rarick and Oyster (1964) noted considerably reduced correlations between skeletal age and the physical fitness variables, diminishing the significance of skeletal age in explaining individual variations in tests of physical fitness. In light of the limited information available concerning the relationships between skeletal age and physical fitness variables in young school-age children, the study described in this chapter was undertaken to determine whether a relationship exists between maturity (in terms of skeletal age) in 6-year-old boys and girls and a series of motor and physical performance tests.

Methods

The study was limited to 6-year-olds: 200 boys and 194 girls. These children belonged to a larger group of subjects from the LEGS-study (Hebbelinck et al., 1980), a longitudinal growth study still in progress. The age range for the 6-year-old children was from 66 to 78 months. Skeletal age was used as an indication of biological maturity. To evaluate skeletal development, an x-ray photograph of the left hand and wrist area was taken, as recommended by Greulich and Pyle (1959). Each radiograph was evaluated by a trained specialist from the laboratory of Human Biometry at the Free University of Brussels (V.U.B.), using the TWII technique (Tanner, Whitehouse, & Healy, 1962).

The physical fitness variables were obtained from the test battery developed by Hebbelinck and Borms (1978) for primary school children. The tests yielded reliability coefficients varying from $.91 < r < .95$. The battery included handgrip strength (DYN), 30 s sit-ups (STU), standing broad jump (SBJ), medicine ball put (MED), 25 m dash (SPR), hockeyball throw (HOC), and a 1-minute run in place (180 st/min) recovery test (REC). A detailed description of the equipment and the test procedures is reported elsewhere (Hebbelinck, 1980; Hebbelinck & Borms, 1978).

An index of relative biological maturity was calculated based on the skeletal age/chronological age ratio (SA/CA-index). Based on a cumulative frequency distribution of the SA/CA-index, three maturity groups were subsequently formed: an early (< 25th percentile), a late (> 75th percentile), and an average (25th percentile $<$ SA/CA $<$ 75th percentile) maturity group. The maurity indices were, respectively, lower than 0.897 for the boys ($n = 48$) and 0.964 for the girls ($n = 47$) in the late maturity group, and 1.170 for the boys ($n = 51$) and 1.104 for the girls ($n = 50$) in the early maturity group. The two extreme groups (i.e., the "early maturers" and the "late maturers") were used for analysis.

Normality of the distributions was tested at the .001 level of significance by the Kolmogorov-Smirnov test for goodness of fit. The null hypothesis for the difference between the means of the two maturity groups was tested by applying a Student t-test at the .05 level of significance, and Pearson product-moment correlations were used to test the significance of the relationships between skeletal age and the physical fitness variables ($p < .05$).

Results

The results of the t-ratios to test the significance of the differences of the physical fitness variables and chronological and skeletal age between the early- and

Table 1. Means, standard deviations, and *t*-ratios of skeletal age and physical fitness variables for early-maturing and late-maturing boys

Variables	Early maturers			Late maturers			
	n	M	SD	n	M	SD	t-ratio
CA (yrs)	51	5.95	.25	48	5.92	.25	.60
SA (yrs)	51	7.49	.58	48	4.80	.45	25.65**
STU (nr)	31	7.81	5.23	27	7.85	5.50	.01
REC (b/min)	17	40.47	3.12	14	39.64	3.93	.65
MED (m)	26	1.75	.44	25	1.47	.36	2.52**
SBJ (cm)	32	95.65	18.65	27	90.44	16.88	1.12
DYN (kg)	32	11.12	3.99	26	9.19	3.43	1.95*
SPR (s)	31	6.50	.52	27	6.70	.63	1.34
HOC (m)	30	7.32	3.86	26	7.61	2.69	.33

Note. CA = chronological age; SA = skeletal age; STU = 30 s sit-up; REC = 15 s recovery pulse rate 15 s after 1-min run; MED = medicine ball put; SBJ = standing broad jump; DYN = handgrip strength; SPR = 25 m dash; HOC = hockeyball throw.
*$p < .05$; **$p < .01$.

late-maturing boys are given in Table 1. Significant values of the *t*-ratios appear in the two strength tests, that is, handgrip strength ($p < .05$) and medicine ball put ($p < .01$). As expected, skeletal age differs significantly between the two groups ($p < .001$), whereas there is no difference in chronological age. Table 2 indicates that besides skeletal age only the medicine ball put ($p < 0.01$) differs significantly between the early- and late-maturing girls, whereas handgrip strength (DYN) just fails to show a significant difference ($p < .05$).

Correlations between strength tests and variables of general physical growth (i.e., height and weight) are usually high. Therefore, in order to rule out the possible effects of height and weight upon the correlations of the measures of physical fitness with skeletal age, partial correlations were computed holding

Table 2. Means, standard deviations, and *t*-ratios of skeletal age and physical fitness variables for early-maturing and late-maturing girls

Variables	Early maturers			Late maturers			
	n	M	SD	n	M	SD	t-ratio
CA (yrs)	50	5.93	.27	47	5.97	.28	.61
SA (yrs)	50	7.18	.64	47	5.11	.60	16.40**
STU (nr)	36	9.50	4.13	34	7.91	5.24	1.42
REC (b/min)	20	41.30	2.81	22	39.73	3.51	1.59
MED (m)	28	1.44	.35	29	1.25	.33	2.14*
SBJ (cm)	35	95.03	16.80	34	88.82	17.24	1.51
DYN (kg)	34	10.53	2.97	34	9.12	3.23	1.88*
SPR (s)	36	6.59	.63	33	6.83	.74	1.47
HOC (m)	35	4.79	1.23	34	4.81	2.23	.07

Note. CA = chronological age; SA = skeletal age; STU = 30 s sit-up; REC = 15 s recovery pulse rate 15 s after 1-min run; MED = medicine ball put; SBJ = standing broad jump; DYN = handgrip strength; SPR = 25 m dash; HOC = hockeyball throw.
*$p < .05$; **$p < .01$.

Table 3. Partial correlations (r) between physical fitness variables and skeletal age of 6-year-old boys

Variables	Zero order correlation	Weight held constant	Height held constant	Height and weight held constant
STU(nr)	.10	.09	.03	.04
REC (b/min)	.05	.00	.01	.02
MED (m)	.36**	.28**	.24**	.23**
SBJ (cm)	.16*	.19*	.11	.13
DYN (kg)	.35**	.21*	.17*	.15*
SPR (s)	– .23*	– .14	– .14	– .13
HOC (m)	.06	.07	.05	.05

Note. STU = 30 s sit-up; REC = 15 s recovery pulse rate 15 s after 1-min run; MED = medicine ball put; SBJ = standing broad jump; DYN = handgrip strength; SPR = 25 m dash; HOC = hockeyball throw.
$*p < .05; **p < .01.$

weight and height, separately or combined, constant. Reference to Table 3 shows that in boys the correlation coefficients are generally low, with the highest values for the correlations between skeletal age and handgrip strength ($r = .35$) and medicine ball put ($r = .36$). By holding weight and height constant, separately or combined, all correlations between skeletal age and the respective physical fitness variables were lowered, leaving only the correlations with medicine ball put and handgrip strength significant at the .05 level of confidence.

Similarly, product-moment correlations were calculated for the girls' sample, holding constant height and weight, separately and combined. Reference to Table 4 shows that all correlations are again very low between skeletal age and the measures of physical fitness. These correlations are reduced by partialing out height and weight, separately or combined. As with the boys, the only significant correlations were found between skeletal age and medicine ball put and handgrip strength. However, when weight is held constant, this significant correlation with the medicine ball put disappears.

Table 4. Partial correlations (r) between physical fitness variables and skeletal age of 6-year-old girls

Variables	Zero order correlation	Weight held constant	Height held constant	Height and weight held constant
STU(nr)	.12	.09	.08	.07
REC (b/min)	.14	.11	.16	.15
MED (m)	.20**	.10	.10	.08
SBJ (cm)	.11	.10	.04	.05
DYN (kg)	.30**	.19**	.15*	.13*
SPR (s)	– .12	– .08	– .06	.05
HOC (m)	.08	.02	.02	.01

Note. STU = 30 s sit-up; REC = 15 s recovery pulse rate 15 s after 1-min run; MED = medicine ball put; SBJ = standing broad jump; DYN = handgrip strength; SPR = 25 m dash; HOC = hockeyball throw.
$*p < .05; **p < .01.$

Discussion

Our hypothesis was that the group with advanced skeletal maturity would show better performance in physical fitness tests than the group with retarded skeletal maturity. The rationale underlying this hypothesis was that the development of strength and motor performance is closely related to skeletal maturity. Indeed, in the early work of Jones (1949) it was shown that early-maturing boys have early-developing strength spurts, and late-maturing boys have late-developing strength spurts. Very little information, however, is available on gross motor performance items, such as running, jumping, throwing, and endurance at the early school age. Although the subjects in the present study were primarily classified according to chronological age, skeletal age differed by more than 2 years (2.69 years in the boys and 2.07 years in the girls) in the groups of early- and late-maturers. The correlations between skeletal age and the two strength measures (handgrip strength and medicine ball put) were significant, both for the boys and the girls. However, when the differences in weight and height were taken into consideration, the significant correlation with the medicine ball put disappeared in the girls.

It is clear that by the age of 6 years a low but significant relationship exists between skeletal maturity and the measures of strength. Although skeletal age gave significant second order partial correlations with the strength measures, it accounts for, at most, 5% of the variance in strength. The modest zero order correlations between skeletal age and the strength items substantiate the relative importance of this maturity factor in 6-year-old boys and girls. This finding corroborates the earlier studies on primary school children of Seils (1951) and Rarick and Oyster (1964) who also reported low, but significant correlations between skeletal age and strength measures and some of the motor performance tasks. When partial correlations were employed, holding height and/or weight constant, little or no correlation between skeletal age and the gross motor skill performance of primary school children remained. Rarick and Oyster (1964) suggested that learning plays a more important role in gross motor performance activities than in tests of static strength, which would explain the higher and statistically significant correlation with strength variables.

In conclusion, the results of this study indicate that skeletal age is of relative importance in accounting for individual differences in the strength of 6-year-old boys and girls. Moreover, height and weight are the factors logically associated with the strength measures. Gross motor performance variables, such as running, jumping, local muscular endurance, and throwing are not significantly related to skeletal maturity status at this age.

References

Asmussen, E., & Heeböl-Nielson, K. (1955). A dimensional analysis of physical performance and growth in boys. *Journal of Applied Physiology, 6*, 585-592.

Greulich, W.W., & Pyle, S.I. (1959). *Radiographic atlas of skeletal development of the hand and wrist*. Stanford: University Press.

Hebbelinck, M., & Borms, J. (1978). *Körperliches Wachstum und Leistungsfähigkeit bei Schulkindern*. Leipzig: Johann Ambrosius Barth Verlag.

Hebbelinck, M., Blommaert, M., Borms, J., Duquet, W., Vajda, A., & Vandermeer, J. (1980). A multidisciplinary longitudinal growth study—Introduction of the project "LLEGS." In M. Ostyn, G. Beunen, & J. Simons (Eds.), *Kinanthropometry II* (pp. 317-325). Baltimore: University Park Press.

Hebbelinck, M. (1980). The research project "Performance and Talent." In J. Simons & R. Renson (Eds.), *Evaluation of motor fitness* (pp. 217-234). Leuven: Institute of Physical Education, KUL.

Jones, H.E. (1949). *Motor performance and growth. A developmental study of static dynamometric strength.* Berkeley: University of California Press.

Malina, R.M. (1975). Anthropometric correlates of strength and motor performance. *Exercise and Sport Sciences Reviews, 3*, 249-272.

Rarick, L.G., & Oyster, M. (1964). Physical maturity, muscular strength, and motor performance of young school-age boys. *Research Quarterly, 35*, 523-531.

Seils, L.G. (1951). The relationship between measures of physical growth and gross motor performance of primary grade school children. *Research Quarterly, 22*, 244-260.

Tanner, J.M., Whitehouse, R.H., & Healy, M.J.R. (1962). *A new system for estimating skeletal maturity from the hand and wrist standards derived from a study of 2600 healthy British children. II. The scoring system.* Paris: Centre International de l'Enfance.

PART III

A Growth and Motor Development Perspective

The papers included in Part III deal with a variety of topics, among them health, fitness, physical performance, shape and proportion, motor ability, motor development, and the effects of training. The theme presentation by Han Kemper outlines the results of a longitudinal growth study carried out in The Netherlands from 1976 to 1979. The study, which focused on the pubescent years (ages 12 through 17), gathered data on fitness and performance, and on health status. The study was especially concerned with changes in fitness-health-performance status occurring during the adolescent years and attempted to relate these to lifestyle changes occurring during these important years. Kemper's report of this extensive investigation is an appropriate introduction to this section. Two of the papers in this section present data on the effects of training on populations well-removed from the average. Borms and his colleagues, in their somatotype analysis of elite body builders, give us some insight into the kind of physical development possible in motivated young men willing to work very hard. In "another part of the forest," Mittleman and her colleagues report on the anthropometric changes produced in older men and women (50 to 77 years) who participated in a 100-day bicycle trip which covered more than 7,000 km. Iida and Matsuura demonstrate some significant relationships between amount of sports involvement and physical performance capacity in their paper on motor performance of Japanese teenagers. Their report leaves unanswered, however, the inevitable question of which (if either) causes the other. Corroll's paper, presenting cross-sectional anthropometric and performance data on Canadian children and youth, may be most useful as a comparative database for other investigators. Another large-scale investigation is that reported by Olgun and Gürses, in which the analysis is based on data from 6,000 Turkish men. Their procedure of dividing somatotypically homogeneous groups into tall, medium, and short subgroups produces some interesting results and may point to a profitable direction for future study.

5

Health and Fitness of Dutch Teenagers: A Review

Han C.G. Kemper
UNIVERSITY OF AMSTERDAM AND FREE UNIVERSITY
AMSTERDAM, THE NETHERLANDS

The review in this chapter is devoted to research on health and fitness of teenagers in The Netherlands. The teenagers of yesterday are the generation of today that brings forth the athletes competing at the Olympic Games in Los Angeles. Longitudinal studies of young people between the ages of 13 and 19 years are seldom found in the literature. Before puberty boys and girls generally show good physical performance capacity in combination with high-activity patterns. At the same time, health authorities complain about the level of physical fitness of youngsters. The teenage period is important in growing towards independence. During these years their lifestyle may change considerably (e.g., food and activity habits) and thus change their health perspective. Individual changes in health and fitness can be described most precisely by studying the same subjects over the time period of interest. Therefore, we designed a longitudinal study to describe the course of development of teenagers in The Netherlands and to find out whether there is a change in health status of this population during growth (Kemper et al., 1983).

Today there is an increasing awareness that the spiraling costs of medical and health service must be controlled and that our society would be better served by devoting more effort to research into methods of primary prevention. This change in attitude has increased the demand for further knowledge about psychosocial factors, influence of environment on health, and the effect of changes in lifestyle. One very important line of research into prevention concerns the early detection of populations at risk and further development of intervention techniques to reduce the probability of morbidity in these populations. The longitudinal approach is most suitable for the elaboration of methods of early detection. In this review, changes in levels of fitness and health of Dutch teenagers observed in our longitudinal study "Growth and Health of Teenagers" will be described and compared with other studies carried out in The Netherlands and other countries.

The design of our multiple longitudinal study (Kemper & van 't Hof, 1978) is a sophisticated compromise between the more traditional approaches to the study of development. In 1976 we started with 307 boys and girls in two clusters: the first and second forms of a secondary school. These boys and girls were from the birth cohorts 1962, 1963, and 1964. The longitudinal measurements were taken during four successive years from 1976 to 1979 (see Figure 1); 233 pupils (131 girls and 102 boys) completed the 4-year study. In this design it is possible to distinguish effects of two confounding factors inherent to longitudinal studies: time of measurement effects and cluster effects.

Although we have only four moments of measurements, having two longitudinal studies which run parallel to each other with a delay of 1 year in age, the study consists of five age groups. This approach makes it possible to estimate a 5-year development in a period of 4 years (Bell, 1954). Additionally, because there is an overlap in age, the two clusters can be compared with each other at three ages. Systematic differences between the two clusters at these three ages is called cluster effect.

At the same time we are able to distinguish a time of measurement effect by comparing children of the same age measured at different years. Another problem that occurs with longitudinal measurements is a testing or learning effect. Many variables—physical as well as psychological—require a certain motivation or habituation of the subjects. This introduces differences between

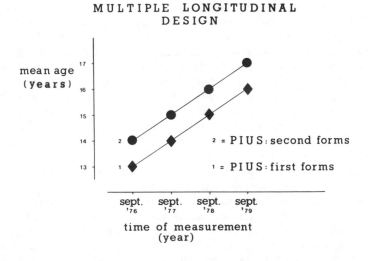

Figure 1. Design of the measurements of the longitudinal group.

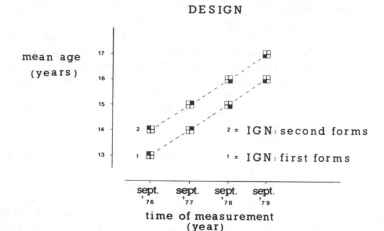

Figure 2. Design of the measurements of the control group.

periods of measuring that are solely due to changes in attitude towards the measurement procedure itself. Such testing effects may be positive (i.e., when learning is important) or negative (i.e., when motivation decreases); the design of our study provides for this. At a second school, comparable with the first, an identical arrangement in clusters was made, but on each measurement occasion a different quarter of the pupils was measured. At the end of the study 292 pupils (159 girls and 133 boys) were measured once each. These measurements are comparable with those of the first school, except that they are not repeated (see Figure 2). Systematic divergence of mean values in the course of the study is an indication for testing effects.

Methods

Physical Fitness

Although the development of an acceptable level of physical fitness is universally accepted as a goal of a physical education program, physical fitness is difficult to define (Clarke, 1967). The President's Council on Physical Fitness and Sport defines it as follows:

> Physical fitness is the ability to carry out daily tasks with vigor and alertness without undue fatigue and with ample energy to enjoy leisure time pursuits and to meet unforeseen emergencies.

As this broad definition suggests, physical fitness can be viewed differently by different professionals. The answer to the definition of physical fitness lies in the domain of construct validity (Alderman & Howell, 1974): On the one hand, it combines expert opinion to define the basic motor abilities based on physiological considerations (Asmussen, 1973; Åstrand & Rodahl, 1977; Margaria, 1966; Shephard, 1968), and on the other hand, factor analytic procedures for identifying the basic motor abilities a test measures (Fleishman, 1964). Physical fitness evaluation through performance testing has been a popular professional pastime for many years. Some of the better known physical fitness test batteries are the AAHPER in the USA (1976), the CAHPER in Canada (1966), the ICSPFT by the International Committee for Standardization of Physical Fitness Tests (Larson, 1974), and in Europe, the EUROFIT by the Council of Europe (1983).

On the basis of a combination of factor analytical procedures (Fleishman, 1964; Simons, Beunen, Renson, & van Gerven, 1982) and physiological considerations (Kemper, 1982), a Motor Performance (MOPER) Fitness Test battery was selected for The Netherlands. This MOPER Fitness Test is a battery of eight simple field tests (see Table 1) to measure strength, speed, flexibility, and aerobic power. In 1977 this MOPER Fitness Test was also administered to a cross-sectional sample of 12- to 18-year-old boys and girls (Bovend'eerdt et al., 1980). The fact that the same tests and the same test procedures were followed in our own longitudinal study gave the opportunity to study test effects that can occur in these fitness tests.

Table 1. The eight test items of the MOPER Fitness Test

	MOPER Fitness Tests	Basic component	Illustration
1	Flexed arm hang	Endurance strength of the arms	
2	10 × 5 m sprint	Running speed	
3	10 leg lifts	Trunk/leg strength	
4	50 plate tapping	Arm speed	
5	Sit and reach	Flexibility	
6	Arm pull	Static arm strength	
7	Standing high jump	Explosive strength	
8	12-min endurance run	Aerobic power	

Health

The definition of health formulated by the World Health Organization (i.e., health is a state of physical, psychological, and social well being) is widely accepted but has not been made operational (Susser, 1973). As a result we chose to describe and measure health as the obverse of disease, illness, and sickness. In this review we will restrict ourselves to the major health problems in our society.

Atherosclerosis is a very important process in the origin of cardiovascular disease. It may begin at a very young age. Study of cardiovascular risk indicators early in life may give us more insight into the etiology of atherosclerotic processes that develop later on in life, and it can also contribute to knowledge about prevention of cardiovascular diseases that are responsible for almost half of all deaths. Since 1975 five studies have been carried out in our country. In each of these, at least three risk indicators were measured: serum cholesterol, blood pressure, and body fat as a percentage of total body weight. Comparison of the results is facilitated because standardized techniques were used: serum cholesterol analyses standardized to the criteria used by the WHO reference laboratories, blood pressure by sphygmomanometer technique, and percentage of fat by skinfold measurements (Durnin & Rahaman, 1967).

According to Cailliet (1979), complaints about back pain occur at least once in the lifetime of more than 80% of all people. In The Netherlands more than 5% of the boys and girls complain about backache (Nijenhuis, 1975). In our longitudinal study the subjects were asked about backache pain each year. Chronic asystematic respiratory abnormalities (CARA) aggravated by environmental pollution confront us with another health problem. We measured forced expiratory volume ($FEV_{1.0}$) related to vital capacity (VC) and expressed it as a percentage: FEV % = ($FEV_{1.0}$/VC) × 100.

Results

Physical Fitness Indicators

Static arm strength, measured by the arm pull test, increased rapidly from age 12 to 17 years in boys and girls (see Figure 3). Until 14 years of age no differences between boys and girls could be demonstrated. The percentile distribution was quite similar to that in Belgian schoolboys (Ostiyn, Simons, Beunen, Renson, & van Gerven, 1982). Comparison of these longitudinal data with the cross-sectional sample revealed a positive testing effect in the longitudinal group. This is in agreement with the study of Ostiyn et al. (1980) who found positive testing effects as well. The reason for this could well be the high motivation for participating in this test. In Figure 4, arm pull is shown in relation to age at peak height velocity (PHV). The rapid increase in arm pull strength during puberty is clearly shown by the S-shaped function in the distance curves. It is also quite evident that the peak in the strength spurt is to be found 1 year after PHV. This is in agreement with the findings of Carron and Bailey (1974). Muscular endurance of the arms, measured by the flexed

ARM PULL

(absolute)

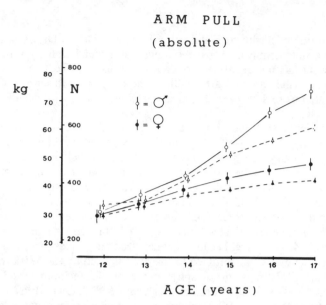

Figure 3. Mean and standard error of arm pull in boys and girls versus calendar age.

ARM PULL

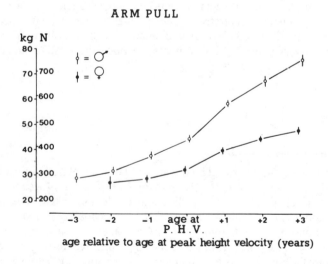

age relative to age at peak height velocity (years)

Figure 4. Mean and standard error of arm pull in boys and girls versus age relative to peak height velocity (PHV).

arm hang, increased in boys but remained constant in girls (Figure 5). Boys showed better performance than girls at all ages. Repeated testing had a rather negative effect on the motivation of the pupils.

Explosive leg strength, measured by the standing high jump, showed no difference between boys and girls until the age of 15 years. From this age on,

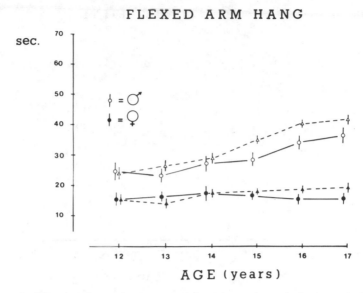

Figure 5. Mean and standard error of flexed arm hang in boys and girls versus calendar age.

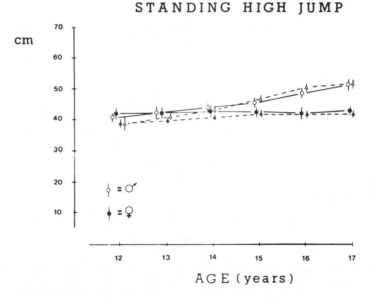

Figure 6. Mean and standard error of standing high jump in boys and girls versus calendar age.

boys became stronger (Figure 6). The trunk/upper leg strength, measured by number of seconds for 10 leg lifts, showed no clear change with age and no difference between the sexes (see Figure 7). In running speed, measured by

10 LEG LIFTS

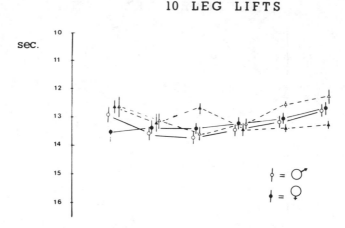

AGE (years)

Figure 7. Mean and standard error of 10 leg lifts in boys and girls versus calendar age.

10 x 5 M SPRINT

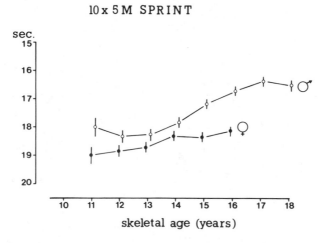

skeletal age (years)

Figure 8. Mean and standard error of 10 × 5 m sprint in boys and girls versus calendar age.

the 10 × 5 m sprint, boys were faster than girls. In both sexes there was a yearly increase in average sprint velocity (Figure 8). Comparison of the longitudinal results with the cross-sectional ones gives no grounds for the assumption of test effects; however, the longitudinal results were almost 1 sec-

50 PLATE TAPPING

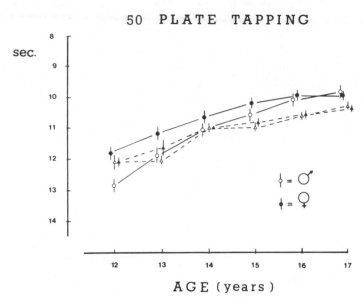

Figure 9. Mean and standard error of 50 × plate tapping in boys and girls versus calendar age.

ond faster than the cross-sectional results. Greater motivation resulting from running in pairs, thus creating an element of competition, could be an explanation for the better scores in the longitudinal group. Arm speed, measured with 50 × plate tapping, showed an increase for both boys and girls (Figure 9). The rate of increase was faster in boys, so that girls, who were superior until the age of 16 years, were equal with boys at ages 16 and 17.

The sit and reach test showed that girls on the average reached 5 cm further than boys. There was an increase with age in both sexes, from the age of 12 to 17 years (Figure 10). The flexibility measured with this test is certainly the result of a complex of different factors that can contribute to "reach" performance, such as compliance of the hamstring muscles, flexibility of the trunk and hip joint, and length of trunk, legs, and arms. The greater trunk height in girls could possibly explain the greater reach compared to boys. When the results are plotted against skeletal age (Figure 11), the spurt in sit and reach performance appears at skeletal ages 12 and 13 in girls and at skeletal ages 14 and 15 in boys. This is also in correspondence with the concomitant increase of trunk height in proportion to leg length (Figure 12).

Aerobic power was measured with the 12-minute endurance run (Figure 13). Boys were superior to girls at all ages and the difference increased with age. There was an indication for a positive testing effect in the boys: the longitudinal group continually increased its performance. Of all the endurance runs, the 12-minute endurance run showed the closest relationship with maximal aerobic power per kg body weight (VO_2max/BW) (Burke, 1979). The correlations, however, varied considerably and were relatively low ($r = .50$ to $.60$). Validation of the 12-minute endurance run performance in our longitudinal sample

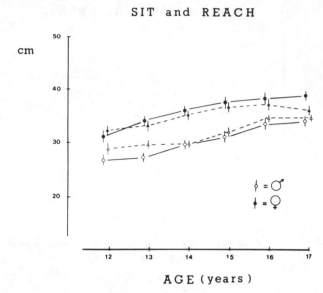

Figure 10. Mean and standard error of sit and reach in boys and girls versus calendar age.

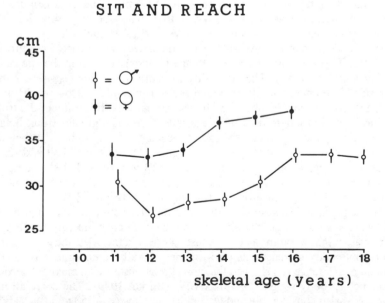

Figure 11. Mean and standard error of sit and reach in boys and girls versus skeletal age.

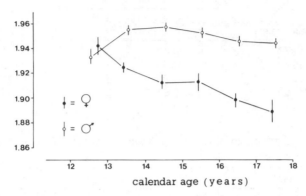

Figure 12. Development of height/sitting height ratio of boys and girls versus calendar age.

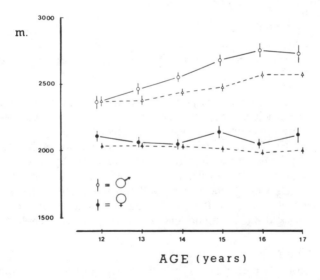

Figure 13. Development of 12-min endurance run of boys and girls versus calendar age.

has been achieved by calculating the correlations with $\dot{V}O_2max/BW$ in the same subjects (see Table 2). The correlations were not very high in boys and even lower in girls. In fact, they explained at least 5% and at the most 36% of the total variation in the prediction of $\dot{V}O_2max/BW$. It can therefore be concluded that the 12-minute endurance run and the $\dot{V}O_2max/BW$ do not measure the same fitness factor.

Table 2. Correlation coefficients between V̇O₂max/BW and 12-min endurance running performance of boys and girls in the 4 years of measurement

Year of measurement	Boys		Girls	
	n	r	n	r
1976/1977	100	.60	118	.33
1977/1978	95	.39	125	.22
1978/1979	96	.51	122	.38
1979/1980	91	.51	101	.39

Table 3. Age effects in physical fitness indicators (MOPER Fitness Test) in boys and girls

Test item	Boys	Girls
Arm pull	↑	↑
Flexed arm hang	↑	=
10 leg lifts	=	=
Standing high jump	↑	=
10 × 5 m sprint	↑	↑
50 plate tapping	↑	↑
Sit and reach	↑	↑
12-min endurance run	↑	=

Note. ↑—increase of mean performance with increase of age; = —mean performance constant with increase of age.

Summarizing the results of physical fitness tests of Dutch teenagers, the following can be concluded: In girls, no decrease in scores with age was found; in four tests there was an increase, and in the other four the performance remained constant (see Table 3). In boys, there was an increase with age in seven out of eight test items. During puberty boys were generally superior to girls over the whole range, except for flexibility (sit and reach) and arm speed (Table 4). Testing effects were found in arm pull (positive) and flexed arm hang (negative) for boys and girls. Positive testing effects were also found in 50 × plate tapping and the 12-minute endurance run in boys (Table 5).

Health Indicators

A high fat mass as a percentage of total body weight (% fat) is considered one of the risk indicators for cardiovascular disease (Winick, 1974). We calculated % fat from the sum of four skinfold thicknesses (Durnin & Rahaman, 1967). In girls the fat mass increased from 23% at age 12 to 28% at age 17. Boys had a mean fat mass of 16% between 12 and 17 years of age (Figure 14). In comparison with pupils from vocational schools, the mean values were 3 to 4% lower (van Beem & Egger, 1979). In Table 6 the percentages of boys and girls that reached these risk values are given per age group. In boys the percentages were low (between 8 and 18%) and did not increase with age. In girls, however, the percentages increased from 10% at age 12 to 28% at

Table 4. Sex effects in physical fitness indicators (MOPER Fitness Test) between age 12 and 17 years

Test item	Sex of higher performance
Arm pull	♂
Flexed arm hang	♂
10 leg lifts	=
Standing high jump	♂ (age 15-16)
10 × 5 m sprint	♂
50 plate tapping	♀ (age 12-14)
Sit and reach	♀
12-min endurance run	♂

Note. ♂—boys higher performance than girls; ♀—girls higher performance than boys; =—no systematic difference between boys and girls.

Table 5. Testing effects of physical fitness indicators (MOPER Fitness Test)

Test item	Boys	Girls
Arm pull	↑	↑
Flexed arm hang	↓	↓
10 leg lifts	—	—
Standing high jump	—	—
10 × 5 m sprint	—	—
50 plate tapping	↑	—
Sit and reach	—	—
12-min endurance run	↑	—

Note. ↑ = increase of mean performance with repeated testing; ↓ = decrease of mean performance with repeated testing; — = no testing effects.

Table 6. Percentage of boys and girls that reach a risk value in body fat per calendar age group

Calendar age group (years)	Body fat < 20%/30%	
	Boys (%)	Girls (%)
12	8	10
13	13	18
14	16	19
15	9	19
16	13	27
17	18	28

age 17, demonstrating that they showed an unmistakable increase of this risk indicator as a mean trend.

Systolic and diastolic (phase V) blood pressure were measured with a sphygomomanometer and standard pressure cuff (12 cm) around the left up-

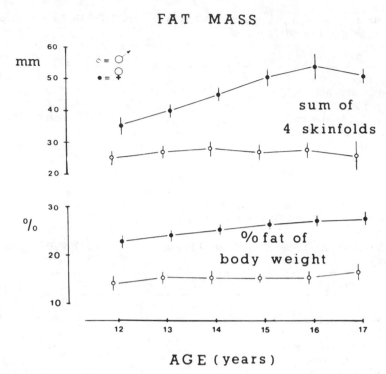

FAT MASS

Figure 14. Mean and standard error of fat mass in boys and girls versus calendar age.

per arm. In Figure 15 systolic (Psyst) and diastolic (Pdiast) blood pressure of boys and girls is shown. Systolic blood pressure in boys increased with age from 16 kPa to 18 kPa. In girls the mean value remained constant at a level of 16 kPa. In the EPOZ study (Hofman, 1983) the same pattern emerged. Part of the increase in Psyst of boys is probably not a genuine rise in pressure but could be traced to a simultaneous increase in total upper arm diameter and the indirect way of measurement with a 12 cm cuff. In comparison with the epidemiological studies in Zoetermeer (Hofman, 1983) and Zutphen (Kromhout, Obermann-de Boer, & Lerenne-Coulander, 1981), the systolic pressures in our study are, on the average, 5 to 10 mmHg higher. In Pdiast, no age effects nor any difference between the two sexes could be established. A drawback of this variable is, however, its low reproducibility demonstrated by the low test-retest correlations in girls of $r = .7$ to $r = .6$, and in boys of $r = .6$ to $r = .5$. We consider a Psyst of more than 19 kPa a risk value for boys and girls in this age period (WHO, 1962). Table 7 shows the percentages of boys and girls that answer to these criteria. In girls the percentage varies between 0 and 3 and does not increase with age; in boys a clearly negative trend is discernible, with percentages increasing from 0% at ages 12/13 to 18% at ages 17/18.

A high total serum cholesterol is closely connected with cardiovascular disease risk. In the etiology of atherosclerosis, a high level of HDL fraction

BLOODPRESSURE

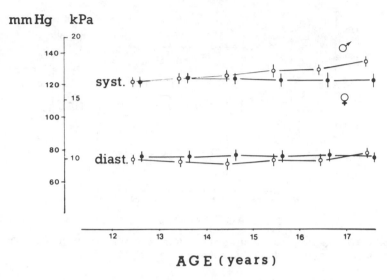

Figure 15. Mean and standard error of systolic and diastolic blood pressure in boys and girls versus calendar age.

is considered a negative risk indicator. In both total serum cholesterol and HDL-cholesterol an interaction between gender and age can be seen at 13 to 14 years (see Figure 16). Before that age boys had higher mean values than girls and after that age lower mean values of total serum cholesterol and HDL-fraction. This phenomenon was also found in the EPOZ study in Zoetermeer. The levels of total cholesterol and HDL-fraction are the same at the corresponding ages compared with the studies in Den Haag (van Gemund, 1980) and Zutphen (Kromhout et al., 1981). The proportion of HDL and total cholesterol is favorable in the case of high HDL-fraction in combination with a low total cholesterol (Arntzenius, 1979). In boys we found a decrease in both total and HDL cholesterol and in girls constant levels of total and HDL cholesterol (lower

Table 7. Percentages of boys and girls reaching a risk value in blood pressure per calendar age group

Calendar age group (years)	Systolic blood pressure > 140 mm Hg	
	Boys (%)	Girls (%)
12	0	3
13	1	1
14	6	2
15	6	3
16	9	0
17	18	3

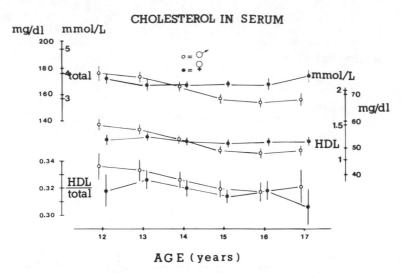

Figure 16. Mean and standard error of total serum cholesterol and HDL-cholesterol in boys and girls versus calendar age.

part of Figure 16). Therefore the proportion between the measures did not change. The mean proportion varied between 0.29 and 0.33 and must be regarded as quite high for these youngsters. If we accept a value of 200 mg/dl of total cholesterol as a risk value for cardiovascular disease (Wilmore & McNamara, 1974), it appears from Table 8 that the percentage of boys and girls reaching this value varied per age group from 1 to 19% in boys and from 3 to 16% in girls. There was, however, no tendency to higher percentages in older age groups. In conclusion, it can be stated that the results showed clear increases in risk indicators for cardiovascular diseases during adolescence, as a mean trend, in systolic blood pressure in boys, and percentage body fat in girls.

The yearly inquiry about pain in the back showed that 19% of the girls and 27% of the boys reported complaints once or more during the 4-year study. About 50% of these pupils had back trouble in only 1 of the 4 years. Figure

Table 8. Percentage of boys and girls that reach a risk value in total cholesterol per calendar age group

| | Total serum cholesterol > 200 mg/dl | |
Calendar age group (years)	Boys (%)	Girls (%)
12	19	3
13	16	13
14	12	12
15	6	12
16	1	11
17	11	16

17 shows the occurrence of back trouble, divided by calendar age groups. These results suggest that the prevalence of back trouble is at its highest point 1 year after reaching the age of peak height velocity (in girls around 13 years and in boys around 14 years). Redistribution of the data in age groups relative to PHV confirms this hypothesis (Figure 18). Fifty percent of the boys and girls who suffered from backache were at the age of PHV or 1 year older. The percentages of boys and girls with backache before the age of PHV were very low: in girls, 1% and in boys, 8%. Two and 3 years after PHV, the percentages were still high: in girls, 42% and in boys, 29%. Consequently, the conclusion is that both height growth velocity and total height are related to complaints of backache.

Vital capacity (VC) of the lungs is determined largely by the dimensions of thorax and lungs. These expand considerably during growth. In boys VC

COMPLAINTS OF BACK

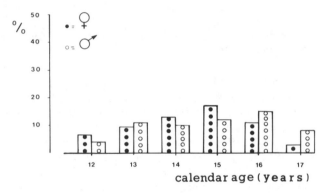

Figure 17. Percentage of boys and girls that have back complaints divided into calendar age groups.

COMPLAINTS OF BACK

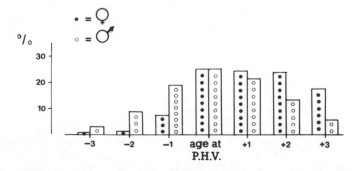

Figure 18. Percentage of boys and girls that have back complaints divided into age groups relative to the age of PHV.

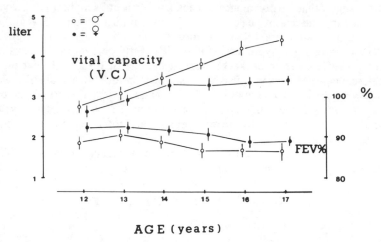

Figure 19. Mean and standard error of vital capacity and FEV % in boys and girls versus calendar age.

increased from 2.5 l at age 12 almost linearly to 4.5 l at age 17. These mean values are in agreement with the norms from Åstrand (1952) and Placheta (1980). The VC in girls increased from 2.5 l to 3.5 l over the same age range (Figure 19). The FEV % is a functional measure that we consider the best estimate of ventilation. FEV % as a mean value amounted to about 92% in girls and to about 88% in boys and did not increase with calendar age. On the contrary, there was a slight tendency to lower percentages as age increased. This, however, is not an indication of a deterioration of the ventilation system but of an increase in body dimensions.

Conclusions

In general, no negative trends in physical fitness indicators could be demonstrated in Dutch teenagers. From the measured health indicators, backache increased in both boys and girls. Clear increases in cardiovascular risk indicators could be demonstrated for systolic blood pressure in boys and for percentage of body fat in girls.

References

Alderman, R.B., & Howell, M.L. (1974). Validity of human performance assessments. In L.A. Larson (Ed.), *Fitness, health and work capacity*. New York: McMillan.

American Association for Health, Physical Education, and Recreation. (1965). *AAHPER youth fitness test manual* (revised ed.). Washington, DC: AAHPER.

Arntzenius, A.C. (1979). Het belang van HDL als negatieve risicofactor en het verband tussen HDL en orale contraceptiva [Importance of HDL as negative risk factor and the relation between HDL and oral contraceptives]. *Nederlands Tijdschrift voor Geneeskuride,* **123**(44), 1910-1912.

Asmussen, E. (1973). Growth in muscular strength and power. In G.L. Rarick (Ed.), *Physical activity, human growth and development.* New York: Academic Press.

Åstrand, P.O. (1952). *Experimental studies of physical working capacity in relation to sex and age.* Copenhagen: Munksgaard.

Åstrand, P.O., & Rodahl, K. (1977). *Textbook of work physiology.* New York: McGraw-Hill.

van Beem, A., & Egger, R.J. (1979). Onderzoek naar de voedingstoestand van 15-16 jarige scholiaren [Research into the nutritional condition of 15-16 year old pupils]. Report of CIVO (TNO) and Nutrition Board. Zeist.

Bell, R.Q. (1954). An experimental test of the accelerated longitudinal approach. *Child Development,* **25**, 281-286.

Bovend'eerdt, J., Bernink, M.J.E., van Hijfte, T., Ritmeester, J.W., Kemper, H.C.G., & Verschuur, R. (1980). De MOPER fitness test, onderzoeksverslag [MOPER fitness test, research project]. Haarlem: De Vrieseborch.

Burke, E.J. (1979). A factor analytic investigation of tests of physical working capacity. *Ergonomics,* **22**(1), 11-18.

Cailliet, R. (1979). *Rugpijn* [Backache]. Lochem: De Tijdstroom.

Canadian Association for Health, Physical Education, and Recreation. (1966). *CAHPER fitness performance test manual.* Ottawa: CAHPER.

Carron, A.V., & Bailey, D.A. (1974). Strength development in boys from 10 through 16 years. *Monographs of the Society for Research in Child Development,* **39**, 4 (Serial no. 157).

Clarke, H.H. (1967). *Application of measurement to health and physical education.* Englewood Cliffs, NJ: Prentice-Hall.

Council of Europe. (1983). *Testing physical fitness. Eurofit, experimental battery, provisional handbook.* Strasbourg: Council of Europe.

Durnin, J.V.G.A., & Rahaman, M.M. (1967). The assessment of the amount of fat in the human body from measurements of skinfold thickness. *British Journal of Nutrition,* **21**, 681-689.

Fleishman, E.A. (1964). *The structure and measurement of physical fitness.* Englewood Cliffs, NJ: Prentice-Hall.

van Gemund, J.J. (1980, 1981). Verloop van bloeddruk en serumcholesterol bij middelbare scholieren [Course of blood pressure and serum cholesterol in secondary school pupils]. In Rapportage Verleende Subsidies. Annual Report NHS, 1980 (pp. 73-74), Annual Report NHS, 1981 (pp. 53-54). Den Haag: Nederlandse Hartstichting.

Hofman, A. (1983). *Blood pressure in childhood.* Unpublished thesis, Erasmus University, Rotterdam.

Kemper, H.C.G. (1982). The MOPER fitness test—a practical approach to standard measurement of custom performances in the field of physical education in The Netherlands. In J. Simons & R. Renson (Eds.), *Evaluation of motor fitness* (pp. 101-114). Leuven: K.U. Leuven.

Kemper, H.C.G., van der Bom, C., Dekker, H., Ootjers, G., Post, B., Snel, J., Splinter, P.G., Storm-van Essen, L., & Verschuur, R. (1983). Growth and health of teenagers in The Netherlands. Survey of multidisciplinary longitudinal studies and comparison with recent results of a Dutch study. *International Journal of Sports Medicine,* **4**, 202-214.

Kemper, H.C.G., & van't Hof, M.A. (1978). Design of a multiple longitudinal study of growth and health in teenagers. *European Journal of Pediatrics,* **129**, 147-155.

Kromhout, D., Obermann-de Boer, G.L., & Lerenne-Coulander, C. (1981). Major CHD risk indicators in Dutch school children, aged 10-14 years. The Zutphen school children study. *Preventive Medicine, 10*(2), 195-210.

Larson, L.A. (Ed.). (1974). *Fitness, health and work capacity. International standards for assessment.* New York: McMillan.

Margaria, A. (1966). Measurement of muscular power (anaerobic). *Journal of Applied Physiology, 21*, 1662-1664.

Nijenhuis, H.W.A. (1975). *Youth employment and health.* Groningen: IRB.

Ostiyn, M., Simons, J., Beunen, G., Renson, R., & van Gerven, D. (1980). *Somatic and motor development of Belgian secondary schoolboys. Norms and standards.* Leuven: Leuven University Press.

Placheta, Z. (1980). *Youth and physical activity.* The development of some functional and morphological indices in 12-15 year old boys with different motor activity. Brno: Purkyne University.

Shephard, R.J. (1968). *Endurance fitness.* Toronto: University of Toronto.

Simons, J., Beunen, G., Renson, R., & van Gerven, D. (1982). Construction of a motor ability test battery for boys and girls aged 12 to 19 years, using factor analysis. In J. Simons & R. Renson (Eds.), *Evaluation of motor fitness* (pp. 151-168). Leuven: K.U. Leuven.

Susser, M. (1973). *Causal thinking in the health sciences, concepts and strategies of epidemiology.* London: Oxford University Press.

World Health Organization. (1962). *Arterial hypertension and ischemic heart disease. Preventive aspects* (Technical Report Series No. 231). Geneva: WHO.

Wilmore, J.H., & McNamara, J.J. (1974). Prevalence of coronary heart disease risk factors in boys 8-12 years of age. *Journal of Pediatrics, 84*, 527-531.

Winick, M. (1974). *Childhood obesity, current concepts in nutrition.* New York: Wiley.

6

Somatotypes of World Class Body Builders

Jan Borms
VRIJE UNIVERSITEIT BRUSSEL
BRUSSELS, BELGIUM

William D. Ross
SIMON FRASER UNIVERSITY
BURNABY, BRITISH COLUMBIA, CANADA

William Duquet
VRIJE UNIVERSITEIT BRUSSEL
BRUSSELS, BELGIUM

J.E. Lindsay Carter
SAN DIEGO STATE UNIVERSITY
SAN DIEGO, CALIFORNIA, USA

Collecting and interpreting physical measurements on humans is one of the aims of kinanthropometry. Olympic and top athletes are a kind of "special" people and have therefore been the focus of several studies, starting as early as 1928 (Knoll, 1928). Although the latest studies on Olympic athletes (Carter, 1982, 1984) provided a considerable amount of new information and introduced new strategies for data analysis, not all questions regarding the characteristics of top athletes were answered. One of the groups of athletes that have been studied only very rarely is body builders. They are a group of individuals who engage in lifting weights for the purpose of increasing muscle definition and improving body configuration and the aesthetic appeal of physique. The effect of their intensive training is a remarkable hypertrophy, which is also to be found in competitive weight lifters and in power weight lifters. However, for the latter athletes, muscle definition is of little athletic or aesthetic concern.

There have been some studies on male body builders (Fahey, Akka, & Rolph, 1975; Kajaba, Zrubák, & Grunt, 1966; Katch, Katch, Moffatt, & Gittleson, 1980; MacDougall, Sale, Elder, & Sutton, 1982; Spitler, Diaz, Horvath, & Wright, 1980; Zrubák, 1970, 1972). However, only a few have focused on somatotypes (Carter, 1978; Hrčka & Zrubák, 1973; Štěpnička, 1974, 1977). Therefore, the invitation to measure top body builders in Cairo was a great opportunity to add new information to the current knowledge regarding the physical characteristics of this type of athlete. The 35th World Amateur Body Building Championships were held in Cairo, Egypt, September 14-19, 1981. Such an event provided a unique opportunity to assemble the best and to investigate them conveniently at one site.[1] In this chapter, the somatotype characteristics of world class amateur body builders will be described and compared by weight category.

Method

Subjects

The subjects in this study were 66 male world class amateur body builders, originating from 27 countries and different ethnic backgrounds. They were measured within 1 to 3 days before competition when presumably the aesthetic qualities being judged were optimal for competitive purposes. At least 50% of the total number of participants were measured. Their participation was voluntary but also depended on several other factors such as time restrictions. Our sample included one gold, two silver, and two bronze medal winners, and several other outstanding performers, comprising a representative group. In the analysis, the sample was subdivided into the four weight categories prescribed by the International Federation of Body Builders (IFBB): light weight ($n = 23$), middle weight ($n = 19$), light-heavy weight ($n = 14$), and heavy weight ($n = 10$).

Procedures

Thirty-five standardized anthropometric measurements were taken on each subject. The measurement techniques were essentially based on those of Martin and Saller (1957), and have been described in detail by Borms, Hebbelinck, Carter, Ross, and Larivière (1979), Carter (1982), and Ross and Marfell-Jones (1983). These measurements were chosen partly to be consistent with the protocol used in the Mexico and Montreal Olympic Games anthropological projects and partly to be consistent with the standardization of the International Working Group on Kinanthropometry (IWGK). The 35 measurements included 9 heights and lengths, 7 breadths, 12 girths, 6 skinfolds, and weight. In addition, 6 lengths were derived from projected heights, and the sum of 6 skinfolds was used to represent the amount of skin and subcutaneous tissue.

[1]The authors are indebted to Mr. Ben Weider, President of the International Federation of Body Builders, for a supportive grant and to Mr. Frank Ego for somatotype photography.

Two experienced examiners took all the measurements with the aid of a trained assistant who also obtained three somatotype photographs. Somatotypes were rated according to the Heath-Carter anthropometric plus photoscopic procedure (Carter, 1980). Inter- and intra-individual reliability of the examiners had been secured on several previous occasions, resulting in correlation coefficients higher than .90. The measurements took place within 3 days prior to the official weighing and competition in a private meeting room of the hotel where participants were staying. The subjects were measured in a relaxed state and not under conditions of tensed or "pumped" musculature.

Informed consent was gained from all subjects prior to measurement. The assembled data for each of the four body builder weight categories were treated in several ways. Normality of the distribution ($p < .01$) was checked with the Kolmogorov Smirnov goodness of fit test. Raw scores were subjected to standard parametric procedures and comparisons between groups made by ANOVA for unequal numbers followed by a Duncan's new multiple range test for determining individual mean differences with significance accepted at the .05 level. Somatotype attitudinal distances (SADs) and means (SAMs) were calculated for each individual and for each group, respectively (Duquet and Hebbelinck, 1977). The significance of the differences between means of the four groups was tested ($p < .05$) using the statistical approach described by Carter, Ross, Duquet, and Aubry (1983).

Results

Table 1 shows the means and standard deviations for age, height, weight, and somatotype of the body builders in the four weight categories, as well as the F-ratios calculated from the one-way analysis of variance. Because the categories are based upon weight (< 70 kg; 70-80 kg; 80-90 kg; and > 90 kg), the mean weights of the four categories differed significantly, as could be expected. Body height was also found to differ significantly among the four categories, heavy weights being the tallest and light weights the smallest. Thus, a higher weight category not only differentiates with regard to weight (and eventually muscle bulk and lateral development) but also height.

The groups did not differ significantly in age, mesomorphy, and ectomorphy. Significant differences were noted only for endomorphy. As expected, the mean values for the body builders emphasized the extreme mesomorphic dominance in their physiques. The highest individual mesomorphy found for a body builder in this study was 11 (a heavy weight), whereas the lowest individual value for mesomophy found was 6 1/2 (one light weight and one light-heavy weight). As a reminder, the average male has a mesomorphy rating in the range from about 3 1/2 to 5. Endomorphy and ectomorphy values were both very low in all four categories.

Individual somatotypes for each weight category were plotted on somatocharts (Figures 1 to 4). The mean somatotypes for each weight category are displayed in Figure 5. The uniquely mesomorphic physique of the body builder is further exemplified in Figure 6 where the means of the body builder groups

Table 1. Means and standard deviations of four body-builder weight categories for age, height, weight, and three somatotype components

Variable	Light (L) (n = 23)	Middle (M) (n = 19)	Light-heavy (LH) (n = 14)	Heavy (H) (n = 10)	F-ratio	Duncan's new multiple range test
Age (years)	30.10 (5.72)	28.65 (6.02)	25.58 (4.45)	29.75 (4.29)	2.20	LH M H L
Height (cm)	164.6 (3.5)	170.9 (4.1)	174.9 (4.1)	183.2 (4.4)	55.33*	L M LH H
Weight (kg)	68.7 (3.4)	78.9 (3.1)	87.5 (5.0)	99.0 (8.2)	112.24*	L M LH H
Endomorphy	1.61 (.34)	1.42 (.25)	1.71 (.43)	1.75 (.35)	2.97*	M L LH H
Mesomorphy	8.35 (.88)	8.71 (.87)	8.96 (1.00)	9.00 (1.08)	1.78	L M LH H
Ectomorphy	1.22 (.47)	1.24 (.42)	1.04 (.13)	1.10 (.21)	1.03	LH H L M

Note. Group means not significantly different ($p < .05$) by post hoc Duncan's new multiple range test are indicated by being joined by underlining.
*$p < .05$.

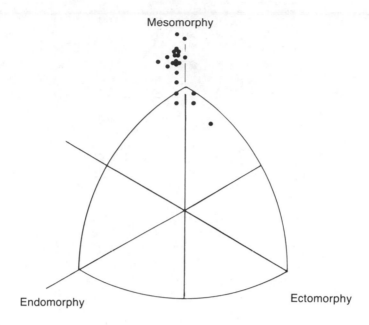

Figure 1. Somatoplots for male light weight body builders (n = 23).

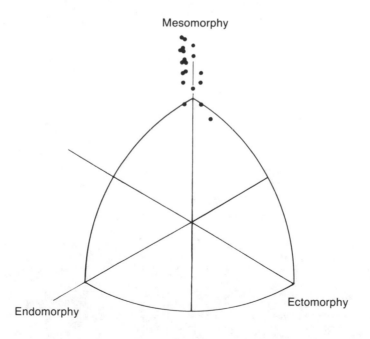

Figure 2. Somatoplots for male middle weight body builders (n = 19).

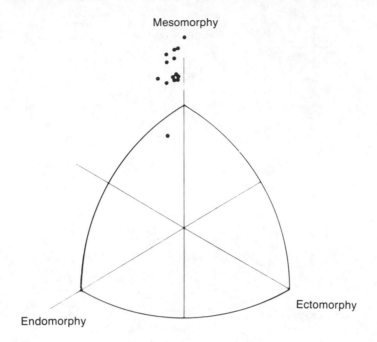

Figure 3. Somatoplots for male light-heavy weight body builders (n = 14).

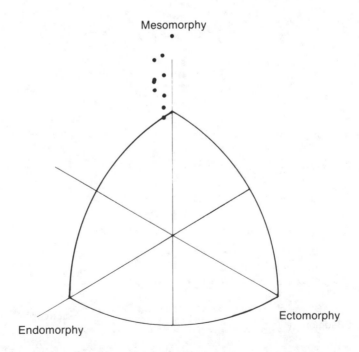

Figure 4. Somatoplots for male heavy weight body builders (n = 10).

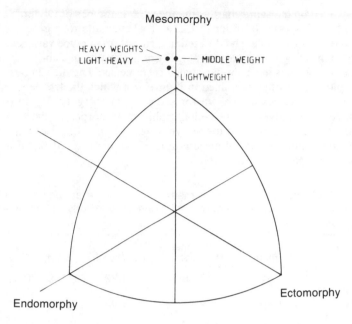

Figure 5. Mean somatoplots for male body builders according to weight category.

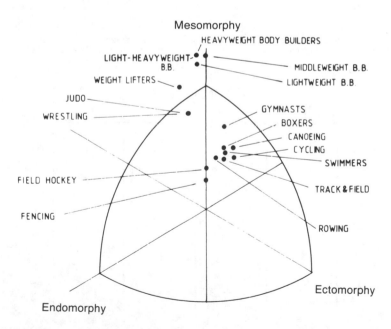

Figure 6. Mean somatoplots for four body building categories and for male Olympic athletes.

are displayed on the same chart as the mean somatotypes of Olympic athletes from different sports. In order to find out whether the four groups differed in somatotype as a whole, and differed in terms of somatotype variance, special tests of significance were carried out. From the results in Table 2, it is clear that the four groups have similar somatotype means. The distributions of the somatoplots are further examined in Figure 7 in which the frequencies of the somatotype categories are shown for each weight category. It can be noted that there are relatively more endomorphic mesomorphs in the two higher weight categories, compared to the two lower weight categories where there are relatively more balanced mesomorphs. The majority (79%) of all competitors are balanced mesomorphs.

Table 2. Somatotype component means, somatotype attitudinal means, attitudinal variance, and standard deviations of four body builder weight categories

| Weight Category | Somatotype means (S) | | | SAM | S_A^2 | S_A |
	Endomorphy	Mesomorphy	Ectomorphy			
Light (n = 23)	1.61	8.35	1.22	0.89	0.30	0.54
Middle (n = 19)	1.42	8.71	1.24	0.86	0.23	0.47
Light-heavy (n = 14)	1.71	8.96	1.04	0.88	0.39	0.62
Heavy (n = 10)	1.75	9.00	1.10	0.93	0.37	0.61

Note. SAM = somatotype attitudinal means, S_A^2 = somatotype attitudinal variance, S_A = somatotype attitudinal standard deviation. No significant difference between means was found.

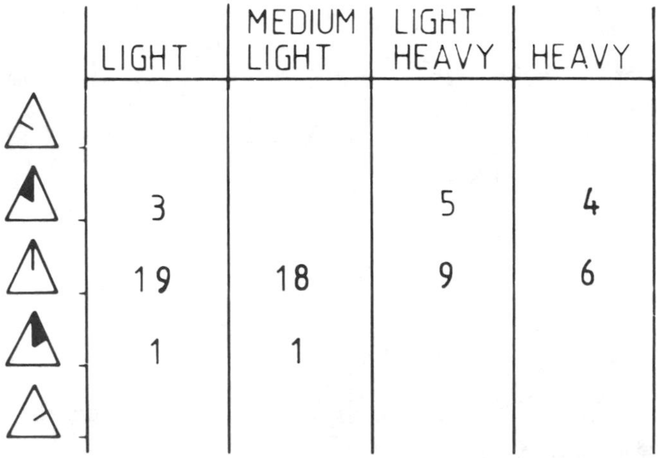

Figure 7. Frequencies per somatotype category and per weight category for male body builders.

Conclusion

One of the objectives of body builders is to achieve maximal muscular hypertrophy regardless of weight class. Body builders differ in height and weight according to weight class, but their somatotypes are very similar. In particular, they have similar mesomorphic components in each class. This indicates that body builders in different weight classes achieve proportionally similar muscular hypertrophy. The findings of this study are similar to those on body builders of Czechoslovakia ($\bar{S} = 2$ 1/2-9-1, Hrčka & Zrubák, 1973; $\bar{S} = 2$-9-1 1/2, Štěpnička, 1974, 1977). However, the present sample is international and is divided into weight categories. Body builders indeed seem to be the prototype of pure extreme mesomorphs, more so than any other group of athletes.

References

Borms, J., Hebbelinck, M., Carter, J.E.L., Ross, W.D., & Larivière, G. (1979). Standardization of basic anthropometry in Olympic athletes—the MOGAP procedure. In V.V. Novotny & S. Titlbachova (Eds.), *Methods of functional anthropometry* (pp. 31-39). Praha: Universitas Carolina Pragensis.

Carter, J.E.L. (1978). Prediction of outstanding athletic ability: The structural perspective. In F. Landry & W.A.R. Orban (Eds.), *Exercise physiology* (pp. 29-42). Miami: Symposia Specialists.

Carter, J.E.L. (1980). *The Heath-Carter somatotype method*. San Diego: San Diego State University Syllabus Service.

Carter, J.E.L. (Ed.). (1982). *Physical structure of Olympic athletes. Part I. The Montreal Olympic Games anthropological project. Vol. 16. Medicine and sport.* Basel: S. Karger.

Carter, J.E.L. (Ed.). (1984). *Physical structure of Olympic athletes. Part II. Kinanthropometry of Olympic athletes. Vol. 18. Medicine and sport science.* Basel: S. Karger.

Carter, J.E.L., Ross, W.D., Duquet, W., & Aubry, S.P. (1983). Advances in somatotype methodology and analysis. *Yearbook of Physical Anthropology, 26*, 193-213.

Duquet, W., & Hebbelinck, M. (1977). Application of the somatotype attitudinal distance to the study of group and individual somatotype status and relations. In O. Eiben (Ed.), *Growth and development: Physique* (pp. 377-384). Budapest: Akademiai Kiado.

Fahey, T.D., Akka, L., & Rolph, R. (1975). Body composition and VO₂max of exceptional weight-trained athletes. *Journal of Applied Physiology, 39*, 559-561.

Hrčka, J., & Zrubák, A. (1973). Telesné zloženie a somatotypy kulturistov, futbalistov a šermiarov [Body composition and somatotypes of body builders, football players and fencers]. *Teorie a praxe telesné výchovy, 21*, 362-368.

Kajaba, I., Zrubák, A., & Grunt, J. (1966). An anthropometric study applied to body builders. *Anthropologie, IV*(3), 19-25.

Katch, V.L., Katch, F.I., Moffatt, R., & Gittleson, M. (1980). Muscular development and lean body weight in body builders and weight lifters. *Medicine and Science in Sport and Exercise, 12*, 330-344.

Knoll, W. (1928). *Die sportärztlichen Ergebnisse der II. Olympischen Winterspiele in St. Moritz 1928.* Bern: Haupt.

Martin, R., & Saller, K. (1957). *Lehrbuch der Anthropologie* (Vol. 1). Stuttgart: Fischer.

MacDougall, J.D., Sale, D.G., Elder, G.C.B., & Sutton, J.R. (1982). Muscle ultrastructural characteristics of elite powerlifters and bodybuilders. *European Journal of Applied Physiology,* **48**, 117-126.

Pipes, T.V. (1979). Physiological characteristics of elite body builders. *The Physician and Sports Medicine,* **7**, 116-120.

Ross, W.D., & Marfell-Jones, M.J. (1983). Kinanthropometry. In J.D. MacDougall, H.A. Wenger, & H.J. Green (Eds.), *Physiological testing of the elite athlete* (pp. 71-115). Ottawa: Canadian Association of Sport Science.

Spitler, D.L., Diaz, F.J., Horvath, S.M., & Wright, J.E. (1980). Body composition and maximal aerobic capacity of body builders. *Journal of Sports Medicine and Physical Fitness,* **20**, 181-188.

Štěpnička, J. (1974). Typologie sportovcu. *Acta Universitatis Carolinae Gymnica,* **1** 67-90.

Štěpnička, J. (1977). Somatotypes of Czechoslovak athletes. In O. Eiben (Ed.), *Growth and development: Physique* (pp. 357-364). Budapest: Akademiai Kiado.

Zrubák, A. (1970). A densiometric study of body builders. *Acta Facultatis Rerum Naturalium Universitatis Comenianae Anthropologia,* **XV**, 89-110.

Zrubák, A. (1972). Body composition and muscle strength of bodybuilders. *Acta Facultatis Rerum Naturalium Universitatis Comenianae Anthropologia,* **XIX**, 135-144.

7

Development of Motor Ability in Senior High School Athletes and Nonathletes

Eio Iida
SHIZUOKA UNIVERSITY
SHIZUOKA CITY, JAPAN

Yoshiyuki Matsuura
UNIVERSITY OF TSUKUBA
SAKURA-MURA, NIIHARI-GUN, JAPAN

Factors such as heredity, living environment, food, and financial aspects of daily life have been established to contribute to the development of motor ability (Inoue & Matsuura, 1976; Kawabata & Hibino, 1961; Ohyama, 1968). A study of their relative importance is indispensable in planning for the optimum development of the motor ability of senior high school students. This study was designed to investigate the relationships between the amount of sport and/or athletic experience and the development of motor ability, and also to what degree the former may contribute to the latter.

Method

A Concept of Fundamental Motor Ability

In this paper, fundamental motor ability was operationally defined as consisting of fundamental motor skills, fundamental motor elements, and physical constitution and functions. This is the hypothesis of motor ability structure of Larson and Yocom (1951). Ability of this type is assumed to be the motor ability element which is essential to any type of motor performance. This concept of fundamental motor ability was defined by Matsuura (1967). It was hypothesized that fundamental motor ability consists of (a) physical constitution, (b) muscular strength, (c) cardiovascular function, (d) muscular endurance, (e) agility, (f) flexibility, (g) circulorespiratory endurance, and (h) fundamental motor skills.

Test Items

Based on the assumptions mentioned above, the following 16 items in eight areas were used to measure the motor ability of the senior high school students in this study:

1. Physical size: stature, sitting height, body weight, and chest girth
2. Muscular strength: back strength and grip strength
3. Cardiovascular function: modified step test
4. Muscular endurance: pull-ups
5. Agility: side step
6. Flexibility: trunk extension and trunk flexion
7. Circulorespiratory endurance: endurance run
8. Fundamental motor skills: 50 m dash, running long jump, vertical jump, and ball throw for distance

The four physical size items were measured during the annual physical examination from April 15 through 25. The remaining 12 test items were administered to the students according to a plan determined by the Ministry of Education, Science, and Culture in the period between late April and early May. In determining the 16 test items, the following principles were taken into consideration:

• The test items should be representative of the subdomains of fundamental motor elements and motor ability mentioned above.
• They should be capable of easy and inexpensive administration to large numbers of subjects without employing special equipment.
• Normative data should be available for comparison and evaluation.
• The test items should be familiar to the students.

Subjects

The data for this paper were collected during a longitudinal survey of students enrolled in K Senior High School in Tokyo, Japan, from 1970 through 1974. This paper refers to only the 12th-grade students. To make the data concerning their sport and athletic club experiences consistent, those who participated as regular club members two or more times a week were selected as club members. Furthermore, their sport and athletic experiences during junior high school were surveyed by questionnaire and their extracurricular activity records from their junior high schools checked to find those who participated two or more times a week. Among those who were selected with the criteria mentioned above, only those whose data on 16 motor ability test items and whose sport and athletic experiences were considered satisfactory were adopted as the samples for this study (Table 1). The subjects were divided into eight groups as follows:

1. Boys with 2 or more years of participation in an athletic club in grades 7 to 9 and grades 10 to 12 as well
2. Boys with 2 or more years of participation in an athletic club in grades 10 to 12
3. Boys with 2 or more years of participation in an athletic club in grades 7 to 9
4. Boys with no experience of participation in an athletic club
5. Girls with 2 or more years of participation in an athletic club in grades 7 to 9 and grades 10 to 12 as well
6. Girls with 2 or more years of participation in an athletic club in grades 10 to 12
7. Girls with 2 or more years of participation in an athletic club in grades 7 to 9
8. Girls with no experience of participation in an athletic club

Construction of an Ability Space

First, a 16 × 16 correlation matrix was computed with the data of all groups pooled. Principal factor analysis procedures were applied to this correlation matrix and the rotated factor pattern matrix was obtained by employing an orthogonal rotation of normal varimax criterion to principal factors corresponding to 1.0 or higher eigen values. Considering the degree of contribution of each factor extracted, the ability space expressed as the factor was constructed with the factors showing the significant degree of contribution to the total variance.

Distribution of Factor Scores

Under the premise that the possible distribution of 16 test variables should be normal, it can be reasonably hypothesized that factor scores which can be obtained from a linear equation of test variables will be normally distributed. Therefore, the probability density function of factor score is

$$y = 1/\sigma\sqrt{2\pi} \cdot e \times p \{-(x - m)^2/2\sigma^2\} \quad ......(1)$$

where m stands for mean factor and σ = standard deviation. Because the extracted factors represent the motor abilities which are measured by the test items used, it becomes possible to investigate the development of motor ability in the orthogonal motor ability space constructed by employing these factors,

Table 1. Athletic club history, group number, and group size for boys and girls

Athletic Club History	Boys Group No.	n	Girls Group No.	n
Two or more years in Group 7-9 and Group 10-12	1	247	5	225
Two or more years in Group 10-12	2	47	6	53
Two or more years in Group 7-9	3	283	7	354
No experience	4	315	8	373
Totals		892		1005

other than in the variable space. When investigating motor ability development, this procedure is more appropriate than the one using items directly from the test, because the space is constructed with the ability axes but not variable ones. In the linear ability space, a distribution curve was drawn in the interval between -3 and $+3$, putting the computed mean and standard deviation into the normal probability density function.

Results and Discussion

The factors extracted and identified via factor analysis were fundamental motor skill, physique and static muscular strength, and flexibility. The degree of contribution of each factor was 43.19%, 16.69%, and 8.86%, respectively (see Table 2). Because Factor 1 (43.19%) was larger in the degree of contribution than the remaining two, this factor was used to investigate the developmental trend in the linear ability space, which was interpreted as fundamental motor skill (see Figure 1).

Comparison of the differences among the eight sample groups revealed the following: For the boys, significant differences ($p < .001$) were found between Groups 1 and 2 (4.545), Groups 1 and 3 (8.716), and Groups 2 and 4 (8.880). Further, a significant difference ($p < .05$) was found between Groups 2 and 3 (2.757). Therefore, Group 1 showed superiority in fundamental motor skill over the other groups, with Group 2 second. In girls, the significant differences ($p < .001$) were found between Groups 5 and 6 (5.355), Groups 5 and 7 (9.848),

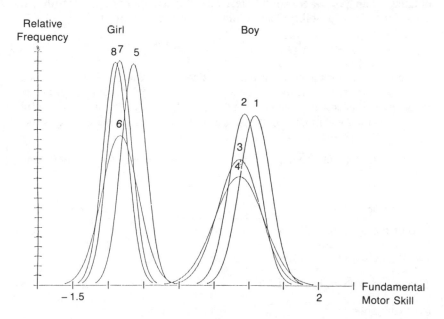

Figure 1. Group configurations in Fundamental Motor Skill axis.

Table 2. Rotated factor matrix for motor ability test items

Item	Factor 1	Factor 2	Factor 3	Communality
1. Stature (cm)	0.73231	0.34924	-0.16360	0.69210
2. Body weight (kg)	0.27412	0.89631	0.08315	0.88543
3. Chest girth (cm)	0.14333	0.88134	0.15639	0.82177
4. Sitting height (cm)	0.66520	0.38511	-0.20840	0.63424
5. Side step (times)	0.74567	0.01644	0.30985	0.65231
6. Vertical jump (cm)	0.87872	0.12892	0.07105	0.79283
7. Back strength (kg)	0.80035	0.37613	0.01236	0.78219
8. Grip strength (kg)	0.79587	0.38644	0.01737	0.73806
9. Modified Harvard Step Test (point)	0.37989	-0.14501	0.26502	0.23558
10. Trunk extension (cm)	0.21322	0.08183	0.64151	0.46369
11. Trunk flexion (cm)	-0.04112	0.09371	0.58015	0.34705
12. 50 m dash (s)	-0.91219	-0.13550	-0.05018	0.85297
13. Running broad jump (cm)	0.91779	0.13232	0.00618	0.85988
14. Handball throw (m)	0.36422	0.28379	0.05456	0.83039
15. Chinning (times)	-0.65859	-0.39064	0.38366	0.73353
16. Endurance run (s)	0.43285	0.44597	-0.49306	0.62936
Amount of contribution	6.91048	2.66960	1.41725	

Groups 5 and 8 (14.911), and Groups 7 and 8 (5.555). Further, a significant difference ($p < .01$) was found between Groups 6 and 8 (2.728). Therefore, Group 5 was superior in fundamental motor skill over the others, followed by Groups 6, 7, and 8. As far as the mean values were concerned, significant differences were observed between these groups. However, their distributions overlapped as shown in Figure 1.

All kinds of physical movements are performed by employing various abilities rather than employing only one specific ability. Therefore, it seems essential to compare the abilities between the sample groups, taking the relationships between their potential abilities and their performances into consideration. From this viewpoint, the two-dimensional ability space was constructed by using two axes which show two abilities out of the three abilities extracted as factors. The horizontal axis represents Fundamental Motor Skill extracted as Factor 1, and the vertical axis Physical Constitution and Muscular Strength extracted as Factor 2.

Table 3 shows the means of the two factors for each of the eight groups, their variances, and covariances. When comparing the mean vectors of these sample groups, a significant difference was observed between Groups 1 and 2, Groups 1 and 3, Groups 1 and 4, Groups 2 and 3, and Groups 2 and 4. By employing a planimeter, the following relative area ratio was obtained for boys:

Groups 1 : 2: 3 : 4 = 1 : 0.8904 : 1.3444 : 1.5060

In girls, the individual areas increase from Group 5 to 8 in that order. Judging from the data, it can be expected that in both boys and girls, 2 or more years

Table 3. Means, variance, and covariance for fundamental motor skill (Factor 1) and physique and strength (Factor 2)

Group	Factor	n	Mean	Variance	Covariance
1	1	247	1.0859	0.22752	− 0.15735
	2		0.24133	0.86154	
2	1	47	0.93548	0.22536	− 0.17092
	2		0.22099	0.72215	
3	1	283	0.85699	0.30716	− 0.41871
	2		0.38793	0.15919	
4	1	315	0.85782	0.35482	− 0.36358
	2		0.34916	0.14553	
5	1	225	− 0.64792	0.17515	− 0.17065
	2		− 0.40588	0.61334	
6	1	53	− 0.83761	0.25852	− 0.26573
	2		− 0.38691	0.69780	
7	1	354	− 0.84150	0.17282	− 0.20706
	2		− 0.22917	0.73594	
8	1	373	− 0.90154	0.17394	− .18364
	2		− 0.25596	0.66504	

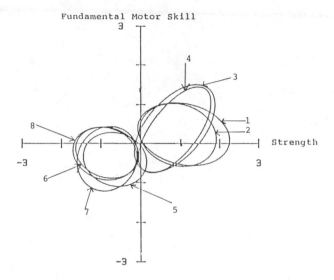

Figure 2. The group configuration in two-dimensional ability space: Fundamental Motor Skill-Physique and Strength Space.

of athletic club participation in the 6 years of grades 7 through 12 would contribute significantly to the development of fundamental motor skills, physical constitution, and static muscular strength. In spite of the above analysis, however, a great deal of overlap occurred in the distributions of the different athletic participation groups (as shown in Figure 2). The significant differences in mean values between different groups only suggest that the groups differ significantly from each other.

References

Inoue, F., & Matsuura, Y. (1976). On the developmental change of factorial structure of motor ability—The tree diagrams of motor ability. *Japanese Journal of Physical Education, 21*(1), 27-37.

Kawabata, A., & Hibino, S. (1961). Studies on growth and motor ability of students in various communities. *Japanese Journal of Physical Education, 5*(4), 124-128.

Larson, L.A., & Yocom, R.D. (1951). *Measurement and evaluation in physical, health, and recreation education.* St. Louis: Mosby.

Matsuura, Y. (1967). *Motor development.* Tokyo: Shoyoshoin.

Ohyama, Y. (1968). The factors contributing to the development of motor ability. *Japanese Journal of Physical Education, 13*(1), 58-65.

8

Status and Changes of Anthropometric and Physical Performance Measures of Manitoba Youth, 6 to 18 Years of Age, Between 1967 and 1980

Victor Corroll
UNIVERSITY OF MANITOBA
WINNIPEG, MANITOBA, CANADA

Methods

The common physical fitness and anthropometric measurements of Manitoba youth 6 to 18 years of age were compared on four sets of cross-sectional data obtained in 1967 (Corroll, LaPage, & Nick, 1969), 1970 (Corroll), 1977 (Gutoski & LaPage), and 1980 (CAHPER). Comparisons were made within sex at each age group. Height and weight data were collected in 1970 (Corroll), 1977 (Gutoski & LaPage), and 1980 (CAHPER). One-minute speed sit-up scores were obtained in all four sets of data, while the flexed arm hang and endurance run data were collected only in 1977 (Gutoski & LaPage) and 1980 (CAHPER). Statistical analysis of the measurements involved the use of the *t*-test ($p < .01$ level) for comparisons between the means of various combinations of the four sets of data described previously.

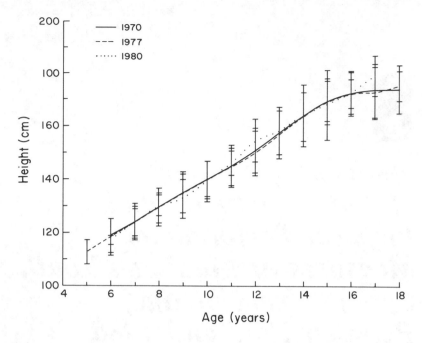

Figure 1. Comparative data (*M, SD*) of standing height for Manitoba males, aged 6 to 18 years.

Results

Height and Weight

In the comparison of the 1970, 1977, and 1980 data, there were no significant height differences (see Figures 1 and 2) within either sex at any age level. The Manitoba height growth pattern, 6 to 17 years of age (Corroll, 1970; Gutoski & LaPage, 1977), was linear with age increase similar to the 1980 Canadian data (Conger, Quinney, Gauthier, & Massicotte, 1982). In the weight comparisons, significant differences appeared as follows (see Figures 3 and 4): The 1980 8-year-old females were heavier than those of 1976, and the 1980 13-year-old females were heavier than those of 1976 and 1970. Also, the Manitoba weight growth pattern, 6 to 17 years of age (Corroll, 1970; Gutoski & LaPage, 1977), was linear with age increase similar to the 1980 Canadian data (Conger et al., 1982).

Examination of the interrelationship of growth (height and weight) with physical performance for both sexes, 6 to 17 years of age, has revealed that physical performance and growth show regular increases in males from 6 to 17 years of age, while the physical performance curves flatten out at the onset of puberty (12 years and over) in females (Conger, Wall, Gauthier, Massicotte, & Quinney, 1982; Corroll, 1969).

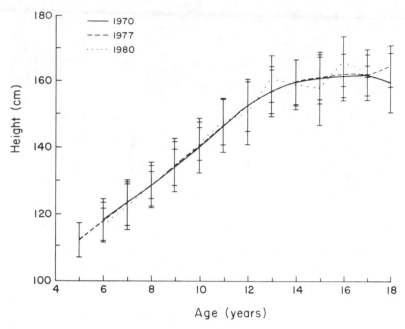

Figure 2. Comparative data (*M, SD*) of standing height for Manitoba females, aged 6 to 18 years.

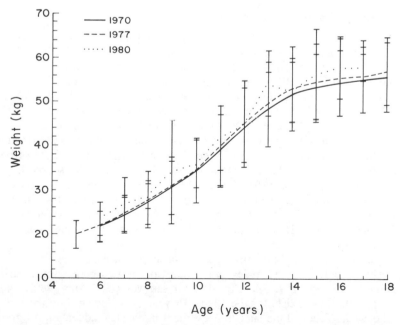

Figure 3. Comparative data (*M, SD*) of body weight for Manitoba males, aged 6 to 18 years.

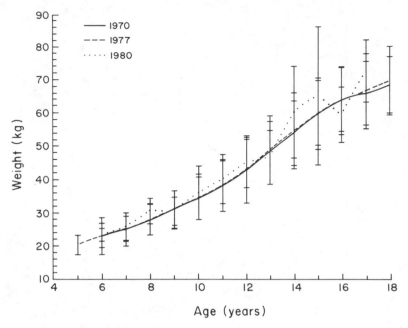

Figure 4. Comparative data (*M, SD*) of body weight for Manitoba females, aged 6 to 18 years.

One-minute Speed Sit-ups

A few statistically significant differences in sit-up performance were shown in various groups for both sexes in the comparison of the 1967, 1970, 1976, and 1980 data (see Figures 5 and 6). These included the following: 6-year-old males and females of 1967 and 1970 scored higher than those of 1976; 7-year-old 1967 males and females scored higher than those of 1970 and 1976; the 1967 and 1970 9- and 10-year-old females scored higher than those of 1977; the 1967 and 1970 11- and 12-year-old males and females scored higher than those of 1977; the 1970 13-year-old males and females scored higher than those of 1967, and the 1977 13-year-old females scored higher than those of both 1967 and 1970; the 1970, 1976, and 1980 14-year-old females scored higher than those of 1967; the 1967 15-year-old males scored higher than those of 1970, but the opposite result occurred for the 15-year-old females; the 1970 and 1977 16- and 17-year-old males and females scored higher than those of 1967; and the 1976 18-year-old males and females scored higher than those of 1967 and 1970.

A general analysis of the differences in sit-up performance scores indicates that the 6- to 12-year-old males and females performed better in the earlier samples (1967 and 1970) than in the later samples (1976 and 1980), while the opposite trend exists for the 13- to 18-year-olds. Insight into the nature of the earlier data and the later improvement in physical education programs may provide some rationale for this opposite trend. The earlier data (1967 and 1970) were collected as part of incentive fitness award programs (Cor-

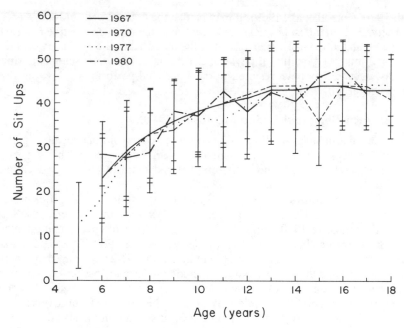

Figure 5. Comparative data (*M, SD*) of 1-minute speed sit-ups for Manitoba males, aged 6 to 18 years.

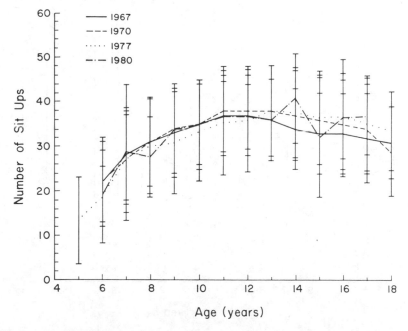

Figure 6. Comparative data (*M, SD*) of 1-minute speed sit-ups for Manitoba females, aged 6 to 18 years.

roll, 1970; Corroll et al., 1969). It may be that the 6- to 12-year-olds are easier to motivate to their best physical performance with tangible incentive rewards (crests) than the 13- to 18-year-olds. On the other hand, the increased contact time and improved quality of physical education in Manitoba schools during the last decade may have resulted in the improved performance level in the older age group.

Flexed Arm Hang and Endurance Run

In the flexed arm hang (see Figures 7 and 8), there were no statistically significant differences within sex at any age level from 6 to 17 years. This suggests that upper body strength should be emphasized more in physical education programs.

In the endurance run (see Figures 9 and 10) the following significant differences were revealed: the 1976 6-year-old boys performed better in the 800 m run than those of 1980, while the 7-, 8-, and 9-year-old males and females of 1980 performed better than their predecessors; the 1980 10- and 11-year-old males and females performed better in the 1600 m run than their predecessors; and the 13-year-old males and females and 15-year-old females performed better in the 2400 m run in 1980. The trend for improved endurance performance in the 1980 sample may also be due to recent improvement in physical education programs in Manitoba (as discussed in the sit-up results).

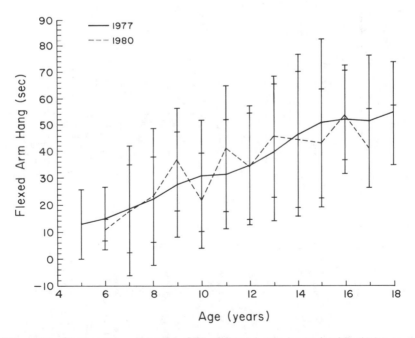

Figure 7. Comparative data (*M, SD*) of flexed arm hang for Manitoba males, aged 6 to 17 years.

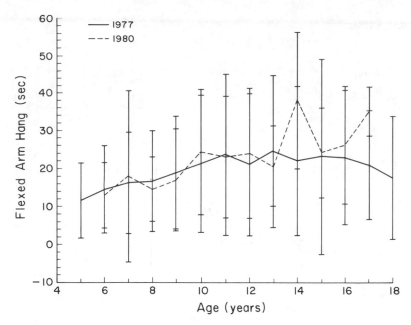

Figure 8. Comparative data (*M, SD*) of flexed arm hang for Manitoba females, aged 6 to 17 years.

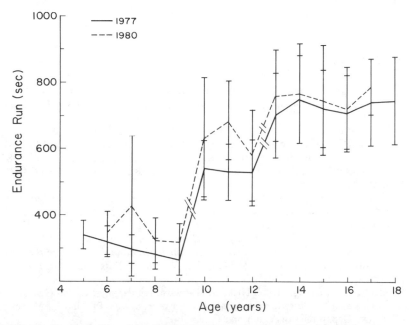

Figure 9. Comparative data (*M, SD*) of endurance run for Manitoba males, aged 6 to 9 years (800 m), 10 to 12 years (1600 m), and 13 to 17 years (2400 m).

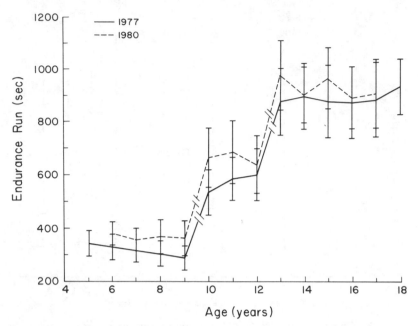

Figure 10. Comparative data (*M, SD*) of endurance run for Manitoba females, aged 6 to 9 years (800 m), 10 to 12 years (1600 m), and 13 to 17 years (2400 m).

References

Canadian Association for Health, Physical Education and Recreation. (1980). *CAHPER fitness performance II test manual, Canadian youth, ages 6-17*. Ottawa: CAHPER.

Conger, P.R., Quinney, H.A., Gauthier, R., & Massicotte, D. (1982). A comparison of the CAHPER fitness-performance test, 1966-1980. *CAHPER Journal,* **48**(6), 6-11.

Conger, P.R., Wall, A.E., Gauthier, R., Massicotte, D., & Quinney, H.A. (1982). Age and sex performance variation of the CAHPER fitness performance II test. *CAHPER Journal,* **48**(6), 12-16.

Corroll, V.A. (1969, October). *A between and within age and sex comparison of physical performance of Manitoba youth, 6 to 18 years of age*. Paper presented at the Canadian Association of Sport Science annual meeting, Calgary, Alberta.

Corroll, V.A. (1970). *Manitoba physical and motor performance and height and weight norms for youth, 6 to 18 years (based on 1970 Manitoba Centennial physical fitness awards plan)*. Province of Manitoba, Winnipeg: Department of Youth and Education.

Corroll, V.A., LaPage, R., & Nick, G. (1969). *Manitoba physical fitness and motor performance manual for Manitoba youth 6 to 18 years of age*. Province of Manitoba, Winnipeg: Department of Youth and Education.

Gutoski, F., & LaPage, R. (1977). *Manitoba physical fitness performance test manual and fitness objectives for Manitoba youth 5-18 years of age*. Winnipeg: Manitoba Department of Education.

9

Anthropometric Changes in Older Cyclists: Effects of a Trans-Canada Bicycle Tour

Karen Mittleman, Susan Crawford, Gordon Bhakthan, Gloria Gutman, and Stephen Holliday
SIMON FRASER UNIVERSITY
BURNABY, BRITISH COLUMBIA, CANADA

The effects of regular aerobic exercise on body composition changes have been studied in middle-aged (Carter & Phillips, 1969; Kukkonen, Rauramaa, Siitonen, & Hanninen, 1982; Lewis et al., 1976; Pollock et al., 1972; Pollock, Cureton, & Greninger, 1969; Pollock, Miller, Linnerud, & Cooper, 1975) and older participants (Adams & deVries, 1973; deVries, 1970; Sidney, Shephard, & Harrison, 1977). Changes observed in these studies have included decreased body weight and skinfold thicknesses, and increased lean body mass. Factors identified as moderators of body composition changes include the intensity, frequency, and duration of the exercise program. Because previous studies have focused on exercise programs typically consisting of moderate aerobic training undertaken 3 or 4 days per week, it is difficult to determine if these results apply to prolonged endurance activity, particularly if undertaken by older adults. Given the present emphasis on the potentially beneficial effects of exercise on older participants, more elderly people are choosing to engage in rigorous and prolonged exercise practices. The gap in the research literature suggests the necessity of assessing the effects of prolonged activity on the elderly population.

The present study examined anthropometric changes experienced by older men and women (50 to 77 years) who completed a 100-day, 7,776 km trans-Canada bicycle tour. The subjects included 16 males (M = 67.3 years, SD

This study was supported by grants from Fitness Canada and Amateur Sports and British Columbia Health Research Foundation.

= 3.7 years) and 6 females (M = 61.5 years, SD = 7.4 years). All subjects were self-selected. Although most participants were not experienced distance cyclists, all had been cycling regularly during a 3-month period prior to the initiation of the study.

Method

Comprehensive anthropometric evaluations were conducted on three occasions: 2 days prior to commencement of the tour (T1), at the midpoint of the tour (T2), and within 2 days of tour completion (T3). Twenty-nine variables including weight, stretched stature, sitting height, 8 skinfold thicknesses, 12 girths, and 6 breadths were measured. Measurements, techniques, and instruments used are described by Ross and Marfell-Jones (1983). Two measures were obtained for each variable and the mean value used for subsequent analyses. If the acceptable tolerance limit (Carter, 1980) was exceeded between the initial measures, a third measurement was taken and the mean of the two closest values was used.

Results

Multivariate analyses of variance (MANOVA) for repeated measures (Hull & Nie, 1981, pp. 32-79) were used to examine changes over the tour and between sexes. Significance was set at $p < .05$. When significant touring effects were found, changes between measurement occasions were examined using standard contrast procedures. Except where noted, all analyses are based on the sample of 16 males and 6 females (Table 1).

Weight and Skinfolds

Significant tour effects were found for body weight. For both males and females, there was a steady weight loss which averaged approximately 3 kg over the tour. A significant gender effect was found, with males having higher values than females. There was no significant change in the sum of eight skinfolds; however, statistically significant tour effects were found for four skinfold measures. A decrease was observed for triceps skinfold thickness, while increases were found in the biceps and medial calf skinfolds. Iliac crest skinfold decreased between times 1 and 2 and increased between times 2 and 3. There were no significant gender by tour interactions for these four skinfold variables.

Significant gender effects were found for triceps, biceps, medial calf, suprailiac, and front thigh skinfolds, with females exhibiting higher values on all variables. Significant gender effects were not observed for subscapular and iliac crest skinfolds. Significant gender by tour interactions were found for suprailiac and front thigh skinfold sites. For front thigh skinfold, females increased throughout the tour, while males increased between times 1 and 2 and

Table 1. Means and standard deviations of weight and skinfold measures for males and females at pre-, mid-, and posttour

Measures	Sex	Pretour		Means and SD Midtour		Posttour	
Weight (kg)[+]	M[a]	70.0	9.8	68.1	8.5	67.5	7.3
	F	61.3	5.8	59.3	6.2	58.0	6.3
Skinfolds (mm):							
Triceps[+]	M	9.4	4.1	8.5	2.3	8.2	2.3
	F	19.1	7.2	18.5	8.8	16.7	6.5
Biceps[+]	M[b]	4.7	1.4	5.9	2.8	6.5	2.7
	F	8.8	4.7	11.0	6.2	10.0	5.1
Iliac crest	M	13.9	6.1	10.9	5.2	12.9	4.9
	F[c]	9.8	4.3	9.3	4.9	10.2	4.5
Suprailiac*	M[b]	6.1	2.5	8.1	4.6	6.7	2.3
	F	13.3	9.1	14.0	9.7	11.7	8.3
Front thigh*	M	11.0	6.0	12.5	7.1	11.4	6.1
	F	25.5	5.4	27.8	6.9	28.6	6.4
Medial calf[+]	M	5.9	2.8	7.3	3.5	7.2	3.0
	F	15.4	4.0	18.5	6.3	17.4	5.6

[a]$n = 15$; [b]$n = 14$; [c]$n = 5$; * = significant sex by time interaction ($p < .05$); [+] = significant sex effect ($p < .05$).

decreased between times 2 and 3. For suprailiac skinfold, females and males increased between times 1 and 2 and decreased between times 2 and 3, but the degree of change was greater for females. No significant gender, tour, or gender by tour effects were found for abdominal and subscapular skinfold thicknesses.

Girths

Statistically significant tour effects were found for seven girth measures. Increases were observed for wrist, relaxed arm, thigh, and ankle girths, while decreases were noted for chest, calf, and neck girths (Table 2). There were no gender by tour interactions for these variables. Skinfold-corrected girths were also used to anthropometrically evaluate the muscular component of lean body mass. When relaxed arm, thigh, chest, and calf girths were corrected for triceps, front thigh, subscapular, and medial calf skinfolds, respectively, significant tour effects were found for arm, chest, and calf girths but not for thigh girth. Significant gender effects were found for arm, chest, and calf girths, with males demonstrating significantly larger values.

Breadths and Heights

Small but statistically significant tour effects were observed for biacromial, biiliocristal, transverse chest breadths, and humerus width (Table 3). A gender by time interaction was found for humerus width, with females decreasing steadily throughout the tour and males demonstrating change only between times 1 and 2.

Table 2. Means and standard deviations of girth measures for males and females at pre-, mid-, and posttour

Girths (cm)	Sex	Pretour		Means and SD Midtour		Posttour	
Wrist[+]	M	16.4	0.5	17.2	0.6	16.9	0.6
	F	15.2	0.5	15.8	0.7	15.5	0.6
Relaxed arm	M[b]	27.5	2.4	28.5	2.0	28.4	1.8
	F	28.1	1.8	28.4	2.2	28.0	2.2
Chest[+]	M	94.6	7.8	93.3	6.9	94.2	5.8
	F[a]	85.7	4.8	84.8	3.7	85.4	4.8
Thigh*	M	51.0	3.0	52.0	3.3	52.5	2.5
	F[a]	53.7	3.7	56.5	2.4	54.5	2.0
Calf	M	35.9	2.7	35.9	2.9	35.4	2.6
	F	34.4	1.5	34.0	1.3	33.5	1.6
Ankle	M	21.2	1.6	21.7	1.9	21.4	1.7
	F	20.0	0.4	20.6	0.8	20.0	0.8
Neck	M	37.3	1.8	37.0	1.8	36.7	1.7
	F	32.2	1.6	32.6	1.4	31.6	1.6
Skinfold corrected girths (cm):							
Arm[+]	M[b]	24.3	1.3	25.7	1.4	25.6	1.3
	F	20.5	1.1	22.0	1.0	21.8	1.1
Chest[+]	M	90.5	6.8	89.4	6.0	90.4	5.3
	F[a]	81.5	3.5	80.1	3.2	80.9	4.0
Calf[+]	M	34.0	2.3	33.4	2.5	33.2	2.2
	F	29.6	1.7	28.2	2.4	28.0	2.2

[a]$n = 5$; [b]$n = 14$; * = significant sex by time interaction ($p < .05$); [+] = significant sex effect ($p < .05$).

Table 3. Means and standard deviations of height and breadth measures for males and females at pre-, mid-, and posttour

Measures	Sex	Pretour		Means and SD Midtour		Posttour	
Height (cm)	M[a]	169.6	4.4	169.6	4.5	170.0	4.5
stretched	F[b]	160.6	8.6	160.7	8.5	161.6	8.5
Breadths (cm):							
Biacromial[+]	M	38.1	1.6	38.0	1.9	38.5	1.4
	F	36.0	2.1	35.1	2.4	35.6	2.8
Transverse chest[+]	M	29.4	1.9	28.5	1.9	29.0	1.7
	F	26.4	2.3	25.1	2.7	25.8	1.9
Biiliocristal	M	29.4	1.7	29.1	1.7	29.2	1.6
	F	30.7	0.9	30.0	0.5	29.9	0.4
Humerus width*	M	7.24	.24	7.18	.29	7.17	.23
	F	6.74	.25	6.65	.26	6.54	.26

[a]$n = 15$; [b]$n = 5$; * = significant sex by time interaction ($p < .05$); [+] = significant sex effect ($p < .05$).

Small but statistically significant tour effects were also found for stretched stature and sitting height. Significant gender effects were found, but no gender by tour interaction. For both males and females there was an increase of 1/2 to 1 cm over the course of the tour.

Discussion

The sexual dimorphic pattern of larger skinfold thicknesses in young adult females versus young adult males was also observed in the older population in the present study. Changes in the individual skinfold measures did not reveal a differential gender effect in the response to training with the exception of the front thigh skinfold, which increased only in the females. The loss of body weight without an accompanying decrease in the sum of skinfold thicknesses measured was somewhat surprising but may be explained in terms of the subjects' exercise history. Previous research has shown that initial body composition is an important factor in determining changes in skinfold thicknesses after training (Brubaker, Kulund, & Evanu, 1978; Pollock et al., 1969). Because the study participants had been training regularly for several months prior to the initial body composition assessment, skinfold changes may have already occurred. This interpretation is supported by the initial sum of six skinfold values. Mean values for both sexes (males, $M = 61.1$ mm, SD = 24.1 mm; females, $M = 107.1$ mm, SD = 38.3 mm) are similar to those reported for active college-age males and females (Yuhasz, 1974).

An alternative explanation for body weight changes without apparent loss of subcutaneous adiposity may be the loss of internal adiposity. Research results have indicated age-related trends toward a redistribution of fat content from subcutaneous to internal adipose tissue (Borkan, Hults, Gerzof, Robbins, & Silbert, 1983; Borkan & Norris, 1977; Skerlj, 1959; Skerlj, Brozek, & Hunt, 1953). This suggests that there is not necessarily a relationship between weight loss and skinfold thickness changes in this older-aged group.

Body composition changes in lean body mass, as evaluated by girth measures, did not follow a consistent pattern. Factor analysis of anthropometric measures have indicated that some girths reflect body fat, others reflect lean body mass, and still others reflect both components of body composition (Hodgdon & Beckett, 1984; Jackson & Pollock, 1976). In the present study, the cycling tour did not produce overall changes in the muscular component of lean body mass determined from skinfold-corrected girths. A trend toward increased arm muscularity was offset by decreases in chest and calf measurements. This finding is not unprecedented. Sidney, Shephard, and Harrison (1977) reported increases in lean body mass (total body potassium) following exercise programs in the elderly; however, these changes were not reflected in circumference measurements. Taken together, the results may indicate that girth measurements are not as meaningful an assessment of lean body mass in an older population as they are in young adults.

The final component of body composition which may be affected by exercise training is bone mass. Bone mass can be evaluated anthropometrically by breadth and girth measures of the wrist and ankle, for example (Jackson

& Pollock, 1976). The declines in breadth measurements we observed during the cycle tour were unexpected. Carter and Phillips (1969) found similar changes in their subjects after habitual aerobic training. Their conclusion that breadth measurements have limited value as indicators of lean body weight changes due to the influence of the underlying soft tissue is in agreement with those of Behnke (1959, 1961). We also feel this may have been a factor in the present study. In addition, blood serum levels of calcium were unchanged throughout the tour, which is an indication that bone mineral loss was not occurring.

Conclusions

Results indicate that aerobic exercise of prolonged duration by older adults does not induce changes in anthropometric measures (specifically in skinfold thicknesses and muscle girths) similar to those reported for middle-aged and young adults. Body weight decline was modest and within the suggested limits for safe weight loss (Williams, 1983). It appears that weight loss with prolonged exercise may not be reflected by losses of subcutaneous adiposity or muscle mass in older individuals. Finally, it is concluded that bicycle touring is an alternative aerobic activity in which older persons, similar to those in the present sample studied, can successfully participate.

References

Adams, G.M., & deVries, H.A. (1973). Physiological effects of an exercise training regimen upon women aged 52 to 79. *Journal of Gerontology, 28*, 50-55.

Behnke, A.R. (1959). The estimation of lean body weight from "skeletal" measurements. *Human Biology, 31*, 295-315.

Behnke, A.R. (1961). Quantitative assessment of body build. *Journal of Applied Physiology, 16*, 960-968.

Borkan, G.A., & Norris, A.H. (1977). Fat redistribution and the changing body dimensions of the adult male. *Human Biology, 49*, 495-514.

Borkan, G.A., Hults, D.E., Gerzof, S.G., Robbins, A.H., & Silbert, C.K. (1983). Age changes in body composition revealed by computer tomography. *Journal of Gerontology, 38*, 673-677.

Brubaker, C.E., Kulund, D.N., & Evanu, K.E. (1978). Cross-country bicycling: Physiological effects. *The Physician and Sportsmedicine, 6*, 74-81.

Carter, J.E.L. (1980). *The Heath-Carter somatotype method* (3rd ed.). San Diego: San Diego State University Syllabus Service.

Carter, J.E.L., & Phillips, W.H. (1969). Structural changes in exercising middle-aged males during a 2-year period. *Journal of Applied Physiology, 27*, 787-794.

deVries, H.A. (1970). Physiological effects of an exercise training regimen upon men aged 52 to 88. *Journal of Gerontology, 25*, 325-336.

Hodgdon, J.A., & Beckett, M.B. (1984). *Prediction of percent body fat for U.S. Navy men from body circumferences and height* (Technical Report No. 84-11). San Diego: Naval Health Research Center.

Hull, C.H., & Nie, N.H. (Eds.). (1981). *SPSS Update 7-9*. New York: McGraw-Hill.

Jackson, A.S., & Pollock, M.L. (1976). Factor analysis and multivariate scaling of anthropometric variables for the assessment of body composition. *Medicine and Science in Sports*, **8**, 196-203.

Kukkonen, K., Rauramaa, R., Siitonen, O., & Hanninen, O. (1982). Physical training of obese middle-aged persons. *Annals of Clinical Research*, **14** (Suppl. 34), 80-85.

Lewis, S., Haskell, W.L., Wood, P.D., Manoogian, N., Bailey, J.E., & Pereira, M. (1976). Effects of physical activity on weight reduction in obese middle-aged women. *American Journal of Clinical Nutrition*, **29**, 151-156.

Pollock, M.L., Broida, J., Kendrick, Z., Miller, H.S., Janeway, R., & Linnerud, A.C. (1972). Effects of training two days per week at different intensities on middle-aged men. *Medicine and Science in Sports*, **4**, 192-197.

Pollock, M.L., Cureton, T.K., & Greninger, L. (1969). Effects of frequency of training on working capacity, cardiovascular function, and body composition of adult men. *Medicine and Science in Sports*, **1**, 70-74.

Pollock, M.L., Miller, H.S., Linnerud, A.C., & Cooper, K.H. (1975). Frequency of training as a determinant for improvement in cardiovascular function and body composition of middle-aged men. *Archives of Physical Medicine and Rehabilitation*, **56**, 141-145.

Ross, W.D., & Marfell-Jones, M.J. (1983). Kinanthropometry. In R. McDougall, H. Wegner, & H. Green (Eds.), *Standards for the physiological assessment of the elite athlete* (pp. 75-115). Ottawa: Canadian Association for the Sport Sciences.

Sidney, K.H., Shephard, R.J., & Harrison, J.E. (1977). Endurance training and body composition of the elderly. *American Journal of Clinical Nutrition*, **30**, 326-333.

Skerlj, B. (1959). Age changes in fat distribution in the female body. *Acta Anatomica*, **38**, 56-63.

Skerlj, B., Brozek, J., & Hunt, E.E. (1953). Subcutaneous fat and age changes in body build and body form in women. *American Journal of Physical Anthropology*, **11**, 577-600.

Williams, M.H. (1983). *Nutrition for fitness and sport*. Dubuque, IA: WIlliam C. Brown.

Yuhasz, M.S. (1974). *Physical fitness manual*. London, Ontario: University of Western Ontario.

10

Relationships Between Somatotypes and Untrained Physical Abilities

Pervin Olgun and Cetin Gürses
TURKISH SPORTS FOUNDATION
ISTANBUL, TURKEY

A study by Ross, Brown, Yu, and Faulkner (1977) examined data on young athletes 12 to 18 years old and showed that the successful ones had somatotypes similar to those of outstanding older athletes. In spite of some interpretational differences, the studies investigating the relationships between somatotype and general motor abilities seem to indicate that mesomorphy is positively associated and endomorphy is negatively associated with physical fitness. Štěpnička (1972, 1976; Štěpnička, Chytrackova, & Kasalicka, 1976) showed that 25 to 60% of the variance in physical fitness tests could be explained by somatotype in adult sportsmen. He also found that somatotype in children was highly related to motor ability test scores.

In our previous study on Turkish adult athletes just before the Mediterranean Games, significant correlations within each sport group were explained by small differences in somatotype (Gürses & Olgun, 1984). We considered that the type, duration, and intensity of training could be maximal within each group at the end of a long training period, and thus in each individual the limits of physical fitness capacity were reached. On the other hand, effects of different types of training (wrestlers compared to volleyball players) were shown to be different on subjects with similar somatotype and height.

These findings, considered together with the finding that (a) somatotype could be changed by physical training (Carter & Phillips, 1969; Carter & Rahe, 1975), and (b) physical fitness could be altered by training, fail to verify the existence of a relationship between somatotype and untrained physical abilities. The aim of this study is to investigate the relationship between somatotypes and untrained physical abilities in men.

Method

Six thousand Turkish males, 18 to 20 years of age, with no special physical training history, were selected for the study. A Modified California test battery was administered for measurement of physical fitness elements. Anthropometric measures necessary for assessing somatotyping according to the Heath-Carter anthropometric somatotyping method (with a height correction for endomorphy) were taken, and somatotypes were calculated by a computer program. For statistical analysis, means, standard deviations, and simple and partial correlation coefficients were calculated. Analysis of variance was applied when appropriate.

Results

Means and standard deviations of physical fitness test scores and somatotype components are given in Table 1. Correlation coefficients, shown in Table 2, do not indicate any statistically significant relationship between physical

Table 1. Means and standard deviations of physical fitness test scores of nontrained males (N = 5840)

Measure	M	SD
Push-ups	18.73	7.8
Broad jump	170.92	26.9
Side steps	15.21	3.5
Sit-ups	25.28	7.0
Chin-ups	4.02	3.4
Shuttle run	327.15	42.3
Endomorphy (ht. cor.)	1.74	0.69
Mesomorphy	4.77	1.21
Ectomorphy	2.62	1.06
Height-weight ratio	12.88	0.54

Table 2. Correlations between physical fitness test scores and somatotype components

Test item	Endomorphy (corrected)	Mesomorphy	Ectomorphy
Push-ups	− .007	.259	− .227
Broad jump	− .096	− .011	.072
Side steps	.029	− .017	.056
Sit-ups	− .013	.084	− .060
Chin-ups	− .118	.071	− .038
Shuttle run	− .007	.025	− .051

fitness test scores and somatotype components, except the slight relationships between push-ups and mesomorphy and ectomorphy.

Another approach to analyzing the effects of somatotype on physical fitness test scores was to classify subjects into somatotype groups defined by somatoplots of Olympic athletes (Figure 1). The mean scores of these groups were analyzed by analysis of variance and no significant difference was observed. This result was thought to be the consequence of the height differences which affect most of the physical fitness test scores. Therefore, the subjects in somatotype groups were subgrouped according to height. Height was defined as "short" for 160-168 cm, "medium" for 169-174 cm, and "tall" for above 175 cm. In Table 3 the numbers of subjects in somatotype-height groups are given. The means of physical fitness tests in "somatotype-height" groups are given in Tables 4, 5, and 6. The comparisons are made between different somatotypes within the height groups and also between heights within somatotype groups. Table 7 indicates which of the differences between height groups within somatotype groups are significant.

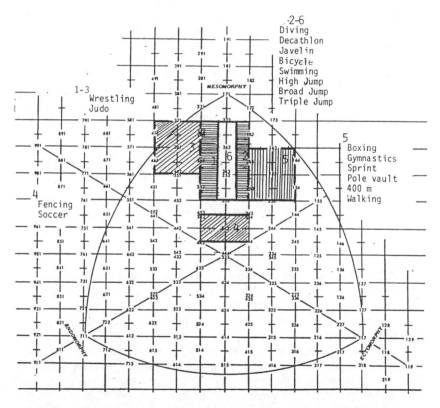

Figure 1. Somatotype groups as defined by somatoplots of Olympic athletes.

Table 3. Number of subjects by somatotype and height groups

Height group	Bal. meso 1	Bal. meso 2	Endo. meso 3	Central 4	Ecto. meso 5	Bal. meso 6	Total
Short	367	541	123	24	428	455	1938
Medium	159	309	39	26	371	231	1135
Tall	49	74	53	69	100	34	379
Total	575	924	215	119	899	720	3452

Table 4. Means and standard deviations of physical fitness test scores of short subjects divided into somatotype groups

Test item	Bal. meso 1	Bal. meso 2	Endo. meso 3	Central 4	Ecto. meso 5	Bal. meso 6
Push-ups	20.52[a] ±7.51	21.03 ±7.78	18.99[a] ±7.38	17.17[a] ±6.30	20.39 ±7.40	20.74 ±7.79
Chin-ups	4.17 ±2.35	4.72 ±2.56	3.50[a] ±1.64	2.92[a] ±2.15	4.79 ±2.74	4.44 ±2.4
Broad jump	168.20 ±22.66	173.22 ±42.36	167.74 ±21.93	164.33 ±24.50	169.79 ±24.39	170.65 ±23.44
Side steps	15.03 ±2.57	15.46 ±5.65	15.05 ±2.38	15.46 ±2.57	14.70 ±2.83	15.31 ±4.64
Sit-ups	25.37 ±6.46	25.64 ±6.59	24.63 ±5.93	25.21 ±6.16	25.77 ±6.26	25.37 ±6.92
Shuttle run	326.51 ±20.10	324.10 ±20.58	326.35 ±19.33	324.08 ±17.44	323.80 ±19.49	326.39 ±26.05

[a]Mean score significantly less than for groups 1, 2, 3, and 4, $p < .05$.

The effect of height on physical fitness tests among subjects within the same somatotype group is summarized as follows: Height affects success in push-up and chin-up tests negatively. This finding of a decrease in mean scores with greater height is observed in all somatotype groups. The decrease in means of push-up and chin-up tests with increased height were found to be significant in most of the comparisons. In broad jump and side step tests the effect of height seems positive in most of the comparisons. Height does not seem related to performance of these subjects in sit-up and shuttle run tests. The following were the effects of somatotype on physical fitness within the height groups. Somatotype groups 1, 2, 5, and 6 seem more successful than the subjects having somatotypes 3 and 4 in push-up and chin-up tests where body weight is carried by the upper extremity and shoulder girdle muscles and also in side step and broad jump tests where the lower extremities are involved.

Table 5. Means and standard deviations of physical fitness test scores of medium height subjects divided into somatotype groups

Test item	Bal. meso 1	Bal. meso 2	Endo. meso 3	Central 4	Ecto. meso 5	Bal. meso 6
Push-ups	17.69 ±6.65	18.68 ±6.80	17.18 ±6.28	16.88 ±8.36	18.77 ±6.68	18.66 ±6.87
Chin-ups	3.11 ±2.24	3.83[c] ±2.35	2.40[a] ±1.23	2.92 ±2.64	4.09[b] ±2.22	3.50 ±2.29
Broad jump	170.72[e] ±24.09	175.73 ±24.82	168.46[d] ±20.59	164.85[e] ±24.00	175.44 ±25.16	174.00 ±24.37
Side steps	16.11 ±6.97	15.74 ±3.01	15.56 ±2.9	16.11 ±2.34	15.45 ±2.72	15.95 ±6.20
Sit-ups	25.24 ±6.58	26.76 ±6.12	26.15 ±6.89	27.04 ±6.65	25.86 ±5.97	25.90 ±6.50
Shuttle run	327.16[g] ±21.63	323.32 ±21.87	336.38[f] ±19.30	326.81 ±24.11	322.80 ±22.08	326.35 ±22.74

[a]Mean score significantly less than for groups 1, 2, 5, and 6, $p < .01$.
[b]Mean score significantly greater than for groups 1, 4, and 6, $p < .01$.
[c]Mean score significantly greater than for group 4, $p < .05$.
[d]Mean score significantly less than for groups 2, 5, and 6, $p < .05$.
[e]Mean score significantly less than for groups 2 and 5, $p < .05$.
[f]Mean score significantly higher than for all other groups, $p < .05$.
[g]Mean score significantly higher than for groups 2 and 5, $p < .05$.

Table 6. Means and standard deviations of physical fitness test scores of tall subjects divided into somatotype groups

Test item	Bal. meso 1	Bal. meso 2	Endo. meso 3	Central 4	Ecto. meso 5	Bal. meso 6
Push-ups[a]	15.79 ±6.88	17.41 ±7.17	8.00 ±3.61	11.33 ±6.41	17.65 ±6.18	16.79 ±6.01
Chin-ups[b]	1.79 ±1.65	3.00 ±2.23	0.67 ±1.15	1.89 ±2.52	3.54 ±2.59	2.35 ±1.84
Broad jump	177.58 ±19.83	183.59 ±21.8	169.00[c] ±6.56	165.78[c] ±22.12	183.54 ±24.69	181.71 ±18.40
Side steps	16.10 ±3.48	16.54[e] ±2.89	14.33[d] ±1.15	15.67 ±1.50	15.97 ±2.93	16.38 ±2.77
Sit-ups	26.05 ±6.22	26.64 ±6.39	21.00[f] ±2.65	24.78 ±7.53	25.91 ±6.85	26.91 ±5.54
Shuttle run[g]	331.89 ±25.59	328.08 ±23.92	333.33 ±14.05	344.33 ±19.77	321.51 ±24.63	326.97 ±26.88

[a]All differences between means are significant except for 2 vs 5, 2 vs 6, 5 vs 6, $p < .05$.
[b]All differences between means are significant except for 1 vs 4, 2 vs 5, 4 vs 6, $p < .01$.
[c]Mean score significantly less than for groups 1, 2, 5, 6, $p < .05$.
[d]Mean score significantly less than for all other groups, $p < .05$.
[e]Mean score significantly greater than for group 1, $p < .05$.
[f]Mean score significantly less than for all other groups, $p < .01$.
[g]All differences between means are significant except for 1 vs 3, 2 vs 6, 3 vs 4, $p < .01$.

Table 7. Significant differences between physical fitness test scores of subjects in the same somatotype group but at different heights

Groups of somatotype category	Short-Medium		Short-Tall		Medium-Tall	
1	Push-ups	$p < 0.001$	Push-ups	$p < 0.001$	Push-ups	$p < 0.05$
	Chin-ups	$p < 0.001$	Chin-ups	$p < 0.001$	Chin-ups	$p < 0.001$
			Broad jump	$p < 0.001$	Broad jump	$p < 0.01$
2	Push-ups	$p < 0.01$	Push-ups	$p < 0.001$	Push-ups	$p < 0.05$
	Chin-ups	$p < 0.001$	Chin-ups	$p < 0.001$	Chin-ups	$p < 0.001$
			Broad jump	$p < 0.01$	Broad jump	$p < 0.001$
			Side steps	$p < 0.01$	Side steps	$p < 0.05$
3	Chin-ups	$p < 0.001$	Push-ups	$p < 0.001$	Push-ups	$p < 0.001$
	Shuttle run	$p < 0.01$	Chin-ups	$p < 0.001$	Chin-ups	$p < 0.001$
			Sit-ups	$p < 0.01$	Sit-ups	$p < 0.001$
			Shuttle run	$p < 0.05$		
5	Push-ups	$p < 0.05$	Push-ups	$p < 0.001$	Chin-ups	$p < 0.05$
	Chin-ups	$p < 0.01$	Chin-ups	$p < 0.001$	Broad jump	$p < 0.01$
	Broad jump	$p < 0.05$	Broad jump	$p < 0.001$		
			Side steps	$p < 0.001$		
6	Push-ups	$p < 0.05$	Push-ups	$p < 0.001$	Push-ups	$p < 0.01$
	Chin-ups	$p < 0.01$	Chin-ups	$p < 0.001$	Chin-ups	$p < 0.01$
			Broad jump	$p < 0.001$	Broad jump	$p < 0.01$
			Side steps	$p < 0.001$		

Note. Mean values appear in Tables 4, 5, and 6.

Discussion and Conclusions

The simple and partial correlation analyses between physical fitness scores and somatotype components failed to reveal any statistically significant relationship between those variables. This result can be explained by the common problem of disassembling somatotype into its components in such analyses. However, the analysis of variance tests applied after clustering the data into somatotype and height groups indicated some significant relations between somatotype-height combinations and untrained physical abilities. In many studies, the main objectives are to show the relationships between somatotype and performance in sports. However, the physical fitness tests used in the present study explain not the performance as a whole but only one part of it.

The degree of correlation between physical fitness test scores and sport performance varies according to the type of sport. For instance, correlations between tests such as sprint, broad jump, and step tests are higher in sprint, broad jump, and distance events. But are these correlations useful in events like gymnastics, volleyball, basketball, soccer, wrestling, and judo where techniques and tactics are the predominant elements of performance? In such events where a wide range of somatotypes exists, the importance of somatotype as a predicting variable can only be as good as the physical fitness tests. With a close examination of these events one can easily see that physical fitness elements are the basic parts of performance. For example, to move the body on apparatus with a maximum speed and power is one of the main elements of performance in gymnastics. This ability can be tested by chin-up, push-up, shuttle run, and broad jump tests effectively.

On the other hand, height is another variable influencing performance with its relation to weight and biomechanical advantage. Taller performers are limited in potential. In this study, the untrained subjects who were similar to Olympic gymnasts in somatotype-height combination achieved the highest scores in the above-mentioned tests. Also, subjects with somatotype-height combinations similar to those of sprinters and broad jumpers performed better in shuttle run and broad jump tests. These examples show the close relationship between somatotype-height combination and inborn physical abilities which are the basic parts of performance.

As to the main elements of performance, more emphasis is given to technique and tactics in the training of beginners. After the performer has reached certain levels of skill in techniques at the beginning, performance can be further developed by increases in physical fitness elements such as speed, power, and strength. For instance, after the preliminary training period, further development of the techniques of spiking and blocking in volleyball is strictly related to abilities in high jump and broad jump which were shown to be related to somatotype-height combination. As seen in the examples given above, development of techniques and tactics is initially of more importance in most of the sports, and this is limited by somatotype. This conclusion explains why top level competitors have a narrow range of somatotype dispersion (Carter, Aubry, & Sleet, 1982).

In conclusion, Carter's suggestion ''to determine suitable physique at the outset which should allow the athlete to concentrate on developing other qualities necessary for top performance" must be remembered. To predict somatotype and performance from preadult ages seems less dangerous than to ignore it because of age-dependent changes in somatotype. As shown in this study, height is one of the most influential factors in performance. Therefore the somatotype-height combination seems the best indicator of performance.

References

Carter, J.E.L., & Phillips, W.H. (1969). Structural changes in exercising middle-aged males during a 2-year period. *Journal of Applied Physiology, 27*, 787-794.

Carter, J.E.L., & Rahe, R.H. (1975). Effects of stressful underwater demolition training on body structure. *Medicine and Science in Sports, 7*, 304-308.

Carter, J.E.L., Aubry, S.P., & Sleet, D.A. (1982). Somatotypes of Montreal Olympic athletes. In J.E.L. Carter, *Physical structure of Olympic athletes: Part I. The Montreal Olympic Games anthropological project: Vol. 16. Medicine and sport* (pp. 53-80). Basel: Karger.

Gürses, C., & Olgun, P. (1984). *Relationship between physical fitness and somatotype in Turkish national athletes*. Istanbul: Turkish Sports Foundation.

Ross, W.D., Brown, S.R., Yu, J.W., & Faulkner, R.A. (1977). Somatotypes of Canadian figure skaters. *Journal of Sports Medicine and Physical Fitness, 17*, 195-205.

Štěpnička, J. (1972). *Typological and motor characteristics of athletes and university students*. Prague: Charles University.

Štěpnička, J. (1976). Somatotype, body posture, motor level and motor activity of youth. *Acta Univ. Carolinae Gymnica, 12*, 1-93.

Štěpnička, J., Chytrackova, J., & Kasalicka, V. (1976). Somatotypic characteristics of the Czechoslovak superior downhill skiers, wrestlers and road cyclists. *Teor. praxe tel Vych., 24*, 150-160.

PART IV

A Sport
Perspective

The papers in this section illustrate the extent to which the techniques of kinanthropometry have been applied to achieve better understanding of the sports performer. The athletes studied participate in sports as diverse as soccer, golf, judo, weightlifting, and squash. Among the three papers primarily concerned with team sports, two investigated the within-sport differences among athletes playing different positions. Soares and his colleagues investigated the Brazilian national basketball team, while Kansal and his co-workers focused on a large group of Indian soccer players. In contrast, Reilly and Bretherton stated that role specificity by position in the sport of field hockey may have been minimized by recent rule changes. Their report considers differences in anthropometry, body composition, and physical performance parameters between elite and lower ranked female field hockey players. Bloomfield and his Australian colleagues are involved in a longitudinal study assessing the development of young athletes in tennis and swimming in comparison to a nonparticipant control group. Among their preadolescent subjects, intergroup differences in size, proportionality, and body composition did not exist. However, the swimmers showed greater strength and flexibility than the tennis players and the reference group. Three of the papers dealt with combative sports. Claessens and his colleagues investigated elite Belgian judoists, compiling extensive anthropometric and physical performance data. Carter and Lucio compared

California high school wrestlers with Olympic wrestlers. In terms of physique and body composition measures, some of the differences between the Olympic and high school performers were surprisingly small compared to within-group differences among the various weight classifications. Rivera and his colleagues present data on Puerto Rican wrestlers varying in age from 6 years to adult. Bale's paper reports physique, body composition, and strength data on 111 sport science students, all of whom were elite athletes currently active in their various sporting endeavors. His report gives further support to the concept of structural specificity in the selection process (either by the performer or by the trainer) for elite sport.

11

Physical Fitness Characteristics of Brazilian National Basketball Team as Related to Game Functions

Jesus Soares, Olga de Castro Mendes,
Calisto Barcha Neto, and
Victor Keihan Rodrigues Matsudo
SÃO PAULO, BRAZIL

The physical fitness characteristics of athletes participating in different sports have been shown in numerous studies (Alexander, 1976; Barnard, 1979; Bell & Rhodes, 1975; Carter, 1970; Macaraeg, 1978; Parnat, Viru, Savi, & Nurmekivi, 1975; Petroski & Duarte, 1983; Di Prampero, Piñera, & Sassi, 1970; Saltin & Åstrand, 1967; Smith & Byrd, 1976; Vaccaro, Clarke, & Wrenn, 1979; Withers, Roberts, & Davies, 1977). General physical characteristics and the specific skill level of basketball players are probably the most important factors that limit the technical and tactical potential of teams during competition.

Some of these characteristics such as height, weight, and age are commonly known and documented in newspapers and sport magazines. However, detailed analyses of physical fitness variables of top basketball players are not common. Some have studied the morphology of these athletes (Muthiah & Sodhi, 1980; Sodhi, 1980; Verma, Mahindroo, & Kansal, 1978), while others have studied physiological aspects (Brown, Moore, Kim, & Phelps, 1974; Parnat et al., 1975; Vaccaro, Wrenn, & Clarke, 1980). Except for Vaccaro et al. (1980), the authors have analyzed physical fitness characteristics of basketball players as a whole and have not considered each function of the game. Therefore, the present study was undertaken to obtain some descriptive data on high-level Brazilian basketball players and to analyze their physical fitness characteristics as related to game functions.

Method

The study was based on 21 Brazilian basketball players (mean age 24.43 ± 3.59 years; mean weight 92.33 ± 14.07 kg; and mean height 197.41 ± 9.70 cm) selected for the 1983 Pan-American Games. The players were divided into three subgroups according to their game function: Group I, centers (n = 7); Group II, forwards (n = 9); and Group III, guards (n = 5). Brazilian National Team players were chosen due to their very successful performance in different championships around the world.

Measures were taken at the Physical Fitness Laboratory in São Caetano do Sul and at the Olympic Center of Training and Research in São Paulo. Physical characteristics were determined by the following anthropometric measures: (a) body weight, (b) standing height, and (c) skinfold measures from the right side (using an average of three measures at each of seven sites: biceps, triceps, subscapular, middle axillary, suprailiac, abdominal, and calf). Maximal oxygen uptake (VO$_2$max) was predicted using the Åstrand and Ryhming (1954) nomogram according to Duarte (1983) on an electromagnetic bicycle ergometer. Motor performance was assessed by the vertical jump test and shuttle run test. The vertical jump was used to measure explosive leg strength (Soares & Sessa, 1983). It was tested allowing arm swing (VJ/wh) and without arm swing (VJ/who). The shuttle run (SR) was used to assess agility and coordination (AAHPERD, 1976; Stanziola & Prado, 1983). These data were analyzed in relation to the players' functions as centers, forwards, and guards. The mean values were compared by one way analysis of variance (ANOVA); differences were analyzed using the Tukey multiple-comparison test at the .05 level.

Results and Discussion

Anthropometric values are presented in Table 1. These data did not show great differences from other national teams which participated in the 1984 Pre-Olympic Games (Table 2). Brazilian players were shorter in stature than only the Canadian players. On the other hand, compared to professional players, they proved to be taller, younger, and heavier than the 1983 NBA Champion Team, the Philadelphia 76ers (Bagatta, 1984) and the Philippines Professional Team (Macaraeg, 1978).

The analysis of data in relation to game functions showed significant differences ($p < .05$) in height and weight but not in skinfold thickness among subgroups. Centers showed the highest values in height and weight followed by forwards and guards. Height is an important feature among basketball players, but its importance is closely related to players' roles in the game. In this study the mean values for height progressively increased from guards to forwards to centers.

Table 1. Brazilian National Basketball Team players—Means, standard deviations, and F-ratios of anthropometric characteristics of the 1983 Pan-American Games

Measure	Centers $n = 7$	Forwards $n = 9$	Guards $n = 5$	M	F
Age (yrs)	23.29[a] ±3.09[b]	25.00 ±3.87	25.00 ±4.06	24.43 ±3.59	—
Height (cm)	206.64 ±4.14	196.91 ±4.58	185.40 ±8.56	197.41 ±9.70	21.08*
Weight (kg)	102.09 ±17.55	91.99 ±6.90	79.28 ±7.33	92.23 ±14.07	5.60*
Skinfolds (7) mean (mm)	7.91 ±2.95	8.30 ±2.17	6.51 ±1.21	7.74 ±2.31	1.00

[a]Means; [b]standard deviations; *$p < .05$.

The mean values of skinfold thickness were lower than those reported by Muthiah and Sodhi (1980) with top-ranking Indian basketball players. The lack of significant differences among subgroups in this study probably reflects the great metabolic demands of extensive and intensive exercise periods on fat deposits.

Maximal oxygen uptake values (mean values in l/min and ml/kg/min^{-1}) are presented in Table 3. The absolute averages for VO$_2$max were found to be high. However, in sports such as basketball where athletes are required to support their own weight, oxygen uptake expressed per kg of body weight is a better indicator of an individual's ability to perform exhaustive work (Åstrand & Rodahl, 1970). When the present data were expressed in this manner (VO$_2$ ml/kg/min), these athletes showed higher values than those reported by other authors (Table 4), but it must be considered that in this study VO$_2$max was determined through an indirect measure in some ways different from the other

Table 2. Means and standard deviations for age, height, and weight of 1984 Pre-Olympic Games basketball teams by country

Country	Age (yrs)	Weight (kg)	Height (cm)
Argentina	—	90.13 ± 8.46	195.08 ± 9.28
Brazil	24.43[a] ± 3.59[b]	92.23 ± 14.07	197.41 ± 9.70
Canada	23.92 ± 2.50	90.83 ± 10.83	198.17 ± 9.20
Cuba	25.67 ± 4.64	—	196.67 ± 8.08
Mexico	23.67 ± 2.71	87.58 ± 4.08	197.08 ± 7.74
Panama	24.50 ± 2.20	—	193.42 ± 5.47
Puerto Rico	25.00 ± 3.64	89.67 ± 12.09	195.00 ± 8.81
Dominican Republic	24.00 ± 2.41	79.75 ± 6.28	195.09 ± 12.27
Uruguay	24.25 ± 2.80	—	194.00 ± 8.63

Note. Reported by Boletim Técnico do Torneio Pre Olimpico de Basketball, São Paulo, 1984.
[a]Means; [b]standard deviations.

Table 3. Means, standard deviations, and F-ratios for maximal oxygen uptake of the 1983 Pan-American Brazilian National Basketball Team players

Unit of measure	Centers $n = 7$	Forwards $n = 9$	Guards $n = 5$	M	F
l/min	6.08[a] ± 0.52[b]	5.52 ± 0.72	6.01 ± 0.60	5.85 ± 0.64	1.68
ml/kg/min^{-1}	59.72 ± 6.92	59.90 ± 5.10	74.36 ± 6.80	63.64 ± 8.84	9.94*

[a]Means; [b]standard deviations; *$p < .05$.

studies. On the other hand, the majority of studies gives results for the group, and the number of centers, forwards, and guards has not been specifically shown. This fact may influence the VO_2max mean values considering that the guards have the highest values in ml/kg/min^{-1}. It was particularly true when the VO_2max mean values were compared among athletes of different positions. Guards showed the highest values (74.36 $\pm$ 6.80 ml/kg/min^{-1}), followed by forwards (59.90 $\pm$ 5.10 ml/kg/min^{-1}), and centers (59.72 $\pm$ 6.92 ml/kg/min^{-1}). Significant differences ($p < .05$) were found between guards and the other two groups.

Centers' and forwards' results were similar to the mean values reported by Vaccaro et al. (1980) for the Maryland University Team. The guards showed values much higher than those in the literature on basketball players. These results may reflect the greater demand for aerobic work at the guard position than in the other positions. Teams that develop strategies relying on constant speed, particularly from their guards, need a higher level of aerobic power than teams that develop their strategies around the center's abilities; this is particularly true for the Brazilian National Basketball Team.

Vertical jump and shuttle run mean values and standard deviations are presented in Table 5. Vertical jumps (47.95 $\pm$ 8.45 cm without arm swing and 61.90 $\pm$ 8.52 cm with arm swing) were lower than those reported by Soares, Duarte, and Matsudo (1981) for the Brazilian National Volleyball Team (53.93 $\pm$ 6.92 cm and 67.71 $\pm$ 7.04 cm). On the other hand, the vertical jump values for these basketball players were 48% (without arm swing) and 69% (with arm swing) higher than those of 18-year-old Brazilian school children (Sessa, Matsudo, & Vívolo, 1978). When values for the vertical jump with arm swing were analyzed according to position, forwards had higher mean scores than centers ($p < .05$). This fact may represent a genuine difference in jumping ability between forwards and centers. It may be confirmed by the lack of significant differences between centers and forwards in scores for vertical jump without arm swing.

Shuttle run mean values in this study (9.67 $\pm$ 0.52 s) were slower than those of the Brazilian National Volleyball Team (9.23 $\pm$ 0.38 s) (Soares et al., 1981). Although the differences among subgroups were not significant, the guards had better mean scores than the others. These results suggest that agility is a very important variable in basketball players, especially for guards.

Table 4. Means and standard deviations of comparative studies on maximal oxygen uptake of basketball players

Team	Maximal oxygen uptake l/min	ml/kg/min^{-1}	Source
Philippines Professional Basketball Team	4.3[a] ± .23[b]	55.70 ± 2.9	Macaraeg, P.V.J. (1978)
South Australian basketball players	4.82 ± .52	58.50 ± 6.1	Withers, R.T. et al. (1977)
1977 University of Maryland basketball team	5.06 ± .66	59.31 ± 6.58	Vaccaro, P. et al. (1980)
Russian basketball players—Tartu University	4.81 ± .16	55.30 ± 1.8	Parnat, J. et al. (1975)
American professional players	—	45.93 —	Parr, R.B. et al., reported by Vaccaro, P. et al. (1980)
American collegiate performers	—	46.30 —	Brown, B.S. et al. (1974)
Brazilian National Basketball Team	5.85 ± .64	63.59 ± 8.8	This study

[a]Means; [b]standard deviations.

Table 5. Means, standard deviations, and F-ratios for the vertical jump and shuttle run of 1983 Pan-American Brazilian National Basketball Team players

Test item	Centers $n = 7$	Forwards $n = 9$	Guards $n = 5$	M	F
VJ/wh (cm)	55.86[a]	66.78	61.60	61.90	4.31*
	±8.05[b]	±8.27	±8.52	±8.52	
VJ/who (cm)	44.14	49.78	50.00	47.95	1.08
	±8.07	±9.48	±6.44	±8.45	
SR (s)	9.99	9.62	9.32	9.67	3.07
	±0.66	±0.38	±0.18	±0.52	

Note. VJ/wh = vertical jump with arm swing; VJ/who = vertical jump without arm swing; SR = shuttle run.
[a]Means; [b]standard deviations; *$p < .05$.

Conclusions

The results contribute to the following tentative conclusions:

* Further studies may focus more attention on the role of each athlete on a basketball team rather than on the team as a whole.
* Aerobic power seems to be important in basketball, particularly for teams which use constant speed as a basic strategy.
* Lower limb muscular strength in forwards and agility in guards are important qualities.
* Lower limb muscular power, aerobic power, and agility are important qualities in athlete selection and in training programs for basketball.

References

Alexander, M.J.L. (1976). The relationship of somatotype and selected anthropometric measures to basketball performance in highly skilled females. *Research Quarterly, 47,* 575-584.

American Alliance for Health, Physical Education, Recreation, and Dance. (1976). *AAHPERD fitness test manual.* Washington, DC: AAHPERD.

Åstrand, P.O., & Rodahl, K. (Eds.). (1970). *Textbook of work physiology.* New York: McGraw-Hill.

Åstrand, P.O., & Ryhming, I. (1954). A nomogram for calculation of aerobic capacity (physical fitness) from pulse rate during submaximal work. *Journal of Applied Physiology, 7,* 218-221.

Bagatta, G. (1984). *American basketball superstar.* Roma: Sponsor.

Barnard, R.J. (1979).Physiological characteristics of sprint and endurance master runners. *Medicine in Science and Sports, 11,* 167-171.

Bell, W., & Rhodes, G. (1975). The morphological characteristics of association football players. *British Journal of Sports Medicine, 9,* 196-200.

Brown, B.S., Moore, G.C., Kim, C.K., & Phelps. R.E. (1974). Physiological and hematological changes among basketball players during pre-season. *Research Quarterly,* **45**, 257-262.

Carter, J.E.L. (1970). The somatotype of athletes—A review. *Human Biology,* **42**, 535-569.

Di Prampero, P.E., Piñera, L.F., & Sassi, G. (1970). Maximal muscular power, aerobic and anaerobic, in 116 athletes performing at the XIXth Olympic Games in Mexico. *Ergonomics,* **13**, 665-674.

Duarte, M.F.S. (1983). Medidas da Potência Aeróbica. In V.K.R. Matsudo (Ed.), *Testes em Ciências do Esporte* (pp. 39-55). São Caetano do Sul: CELAFISCS.

Macaraeg, P.V.J. (1978). A fitness study of a professional basketball team. *Proceedings of the XXI World Congress in Sports Medicine* (p. 218). Brasília: International Federation of Sports Medicine.

Muthiah, C.M., & Sodhi, H.S. (1980). The effect of training on some morphological parameters of top ranking Indian basketball players. *Journal of Sports Medicine,* **20**, 405-412.

Parnat, J., Viru, A., Savi, T., & Nurmekivi, A. (1975). Indices of aerobic work capacity and cardiovascular response during exercise in athletes specializing in different events. *Journal of Sports Medicine,* **15**, 100-105.

Petroski, E.L., & Duarte, M.F.S. (1983). Aptidão Física de Remadores Brasileiros. *Revista Brasileira de Ciências do Esporte,* **4**, 30-39.

Saltin, B., & Åstrand, P.O. (1967). Maximal oxygen uptake in athletes. *Journal of Applied Physiology,* **23**, 353-358.

Sessa, M., Matsudo, V.K.R., & Vívolo, M.A. (1978). Strength development of inferior limbs in students aged 7-18 according to sex, age, weight, height and physical activity. *Proceedings of the XXI World Congress in Sports Medicine* (p. 201). Brasília: International Federation of Sports Medicine.

Smith, D.P., & Byrd, R.J. (1976). Body composition, pulmonary function and maximal oxygen consumption of college football players. *Journal of Sports Medicine,* **16**, 301-306.

Soares, J., Duarte, C.R., & Matsudo, V.K.R. (1981). Perfil de Volibolistas do Centro Olímpico de Treinamento e Pesquisa-SP. *Revista Brasileira de Ciências do Esporte,* **3**(1), 51.

Soares, J., & Sessa, M. (1983). Medidas da Forca Muscular. In V.K.R. Matsudo (Ed.), *Testes em Ciências do Esporte* (pp. 57-68). Sâo Caetano do Sul: CELAFISCS.

Sodhi, H.S. (1980). A study of morphology and body composition of Indian basketball players. *Journal of Sports Medicine,* **20**, 413-420.

Stanziola, L., & Prado, J.F. (1983). Medidas da Agilidade. In V.K.R. Matsudo (Ed.), *Testes em Ciências do Esporte* (pp. 73-77). Sâo Caetano do Sul: CELAFISCS.

Vaccaro, P., Clarke, D.H., & Wrenn, P.J. (1979). Physiological profiles of elite women basketball players. *Journal of Sports Medicine,* **19**, 45-54.

Vaccaro, P., Wrenn, P.J., & Clarke, D.H. (1980). Selected aspects of pulmonary function and maximal oxygen uptake of elite college basketball players. *Journal of Sports Medicine,* **20**, 103-108.

Verma, S.K., Mahindroo, S.R., & Kansal, D.K. (1978). Effect of four weeks of hard physical training on certain physiological and morphological parameters of basketball players. *Journal of Sports Medicine,* **18**, 379-384.

Withers, R.T., Roberts, R.G.D., & Davies, G.J. (1977). The maximum aerobic power, anaerobic power and body composition of South Australian male representatives in athletics, basketball, field hockey and soccer. *Journal of Sports Medicine,* **17**, 391-400.

12

Multivariate Analysis of Fitness of Female Field Hockey Players

Thomas Reilly and Sandra Bretherton
LIVERPOOL POLYTECHNIC
LIVERPOOL, ENGLAND

This study is concerned with field hockey, a sport that makes demands on aerobic and anaerobic mechanisms, flexibility, and many unique skills. Observations on female field hockey players have shown differences in fitness profiles from volleyball specialists (Maksud, Canninstra, & Dublinski, 1976), basketball players, and nonathletes of a similar age (Johnston & Watson, 1968). A more appropriate reference group for high-caliber players may be players with experience in the game at a lower standard rather than other sport specialists or sedentary females. Measures of VO_2max of international players (Withers & Roberts, 1981; Zeldis, Morganroth, & Rubler, 1978) show superior values to those of collegiate players (Maksud et al., 1976). Besides, physique and muscle function have been shown to differ between playing positions (Johnston & Watson, 1968; Verma, Mohindroo, & Kansal, 1979). The changes in field hockey playing rules for women introduced in the 1981-82 season caused increased demands for fitness development and interchangeability of positional roles, possibly rendering previous studies obsolete. The aims of this study were to identify some components that distinguish elite female hockey players from their less proficient counterparts and to examine differences in fitness factors among playing positions.

Method

The sample of 24 players contained elite (mean age = 22.7 ± 2.8 years) and county standard (mean age = 24.1 ± 3.7 years) players in equal numbers.

135

The elite players were drawn from the "Centre of Excellence" squad based at Liverpool, while the county standard group was recruited from club players in Merseyside. Each subject undertook a test battery which consisted of five domains: field tests, aerobic fitness, anthropometry, lung function, and muscular strength and power. Test administration was controlled for time of day: field tests, early afternoon; laboratory tests, early evening. Brief details of the tests follow:

Anthropometry

Height was measured using a stadiometer mounted on an Avery beam balance scale used for determining subjects' body weight. Linearity was computed from these two measures. Skinfold thicknesses were obtained with a Harpenden caliper at the four sites used for estimating the percent body fat according to Durnin and Womersley (1974). Epicondylar widths were measured with an anthropometer (Holtain Ltd), and muscle girths and calf skinfold were determined for computing the somatotype according to Heath and Carter (1967). The sit-and-reach test was used to measure flexibility in cm and was included in this domain.

Muscular Strength and Power

The standing broad jump (SBJ) and vertical jump were used as indicators of explosive strength. Grip strength of right and left hands and the strength of the knee extensors were obtained using Takei Kiki Kogyo dynamometers. In all cases the best of three trials was recorded. Anaerobic power was measured using a stair-run test, the mean of the three best of five trials being recorded.

Lung Function

A dry spirometer (Vitalograph Ltd) was used to measure vital capacity (VC). The FEV_1 and FEV_1 % VC were also calculated. The percentage above normal values for FEV_1 and VC was calculated by reference to the nomogram of Vitalograph Ltd.

Aerobic Fitness

Subjects performed a 12-min graded exercise test on a Monark friction-braked cycle ergometer, the test involving three separate submaximal work loads for 4 min each. The initial work rate was determined in a preliminary investigation and the increments were 150 kpm/min^{-1} (25 W). Heart rate was calculated from a 10 s sample at the end of the 4th minute of each load using ECG. The physical working capacity (PWC_{170}) was calculated by regression analysis of the three submaximal heart rate values and the VO_2max was estimated from the intermediate value (Åstrand & Ryhming, 1954) with correction for the maximal heart rate. The maximal heart rate was obtained from a graded test to exhaustion immediately following the third 4-min workload, the heart rate being monitored over the last 10 s of each minute of exercise.

Field Tests

Field tests consisted of a sprint, a "T"-run dribbling test, and a distance and accuracy skill test. The sprint over 50 (45.45 m) yards was timed to 0.01 s with a Seiko Quartz Sports 100 stopwatch, the fastest of three trials being recorded. The T-run was over 60 (54.55 m) yards while dribbling a leather hockey ball around skittles. The test involved as many circuits of the T-shaped course as possible in 2 min. All subjects were practiced in the drill which excluded use of reversed sticks, and the best of three trials was recorded. The distance and accuracy test involved a combination of dribbling a ball and hitting it at a target, a set sequence being repeated as often as possible within 2 min. Distance traveled was measured to the nearest 2.5 (2.27 m) yards and relative accuracy was calculated by expressing the number of accurate shots as a percent of the number of hits. All subjects were familiarized with the drill before testing took place.

Statistical Analysis

A principal components analysis was performed on each of the five test domains according to the guidelines of Child (1970). Correlations between principal components were also calculated. The principal component scores were utilized in subsequent multivariate analyses of variance (MANOVA). Two MANOVA tests were performed: first, to distinguish between the two levels of playing proficiency; second, to distinguish between the positional roles of forwards, halves, and backs in the whole sample.

Results

The subjects as a whole were close to normal values for percent body fat and physique but were supranormal in lung function, physical work capacity, and estimated VO_2max (Tables 1 and 2). The most pronounced differences among the groups were in the field tests, the cycle ergometer tests, and muscular strength and power. Altogether 10 components that accounted for an average 80.6% of the variance were extracted from the complete battery.

In the first domain the three components extracted were Body Fatness, Linearity, and Body Weight. These accounted for 30%, 27%, and 21% of the variance. Analysis of the muscular strength and power domain produced the two components, Anaerobic Power and Muscle Strength, which respectively explained 68% and 15% of the variance in the domain. Two components in the next domain, Lung Capacity and Lung Power, accounted respectively for 64% and 28% of the variance. The only component extracted in the next domain was Aerobic Fitness, which accounted for 73% of the variance. Two components of the field test data, Dribbling Speed and Accuracy, accounted for 47.5% and 29% of the variance, respectively. The sprint time, T-run time, and distance covered in the distance and accuracy test all had significant latent vectors in the Dribbling Speed component.

Correlation analysis showed that Dribbling Speed was related significantly to Aerobic Fitness ($r = .476; p < .05$), Body Weight ($r = -.427; p < .05$), and Anaerobic Power ($r = .694; p < .001$). Anaerobic Power was related to Aerobic Fitness ($r = -.437; p < .05$) and Body Weight ($r = .678; p < .001$), which in turn was correlated with Lung Capacity ($r = .519; p < .01$).

Table 1. Means and standard deviations of variables in the anthropometry and muscular strength and power domains for female field hockey players

Variables	Elite players n = 12	County players n = 12	Whole sample n = 24
Anthropometry			
Age (yrs)	22.7[a] ± 2.8[b]	24.1 ± 3.7	23.4 ± 3.3
Height (cm)	164.3 ± 5.6	161.3 ± 4.7	162.8 ± 5.3
Weight (kg)	60.6 ± 3.8	59.5 ± 5.2	60.0 ± 4.5
Body fat (%)	23.0 ± 1.9	22.9 ± 2.9	23.0 ± 2.4
Endomorphy	3.3 ± 0.6	3.2 ± 0.8	3.2 ± 0.7
Mesomorphy	3.6 ± 0.5	3.8 ± 0.5	3.7 ± 0.5
Ectomorphy	2.7 ± 0.8	2.3 ± 0.9	2.5 ± 0.9
Flexibility (cm)	24.8 ± 6.0	20.6 ± 6.8	22.8 ± 6.5
Muscular strength or power			
SBJ (cm)	200.2 ± 17.9	179.9 ± 17.3	190.0 ± 20.1
Vertical jump (cm)	40.3 ± 6.0	36.6 ± 4.2	38.5 ± 5.4
Knee extension strength (kg)	98.0 ± 26.6	89.8 ± 15.4	93.9 ± 21.7
Grip strength left (kg)	35.1 ± 5.9	33.0 ± 3.5	34.1 ± 4.3
Grip strength right (kg)	38.2 ± 3.9	35.2 ± 4.3	36.7 ± 4.3
Anaerobic power (kgm/s^{-1})	84.1 ± 10.6	74.5 ± 9.6	79.3 ± 11.0
Stair run time (ms)	466 ± 47	514 ± 39	490 ± 49
Vertical velocity (m/s^{-1})	1.39 ± 0.13	1.25 ± 0.10	1.32 ± 0.13

[a]Means; [b]standard deviations.

Table 2. Means and standard deviations of variables in the lung function, aerobic fitness, and field test domains for female field hockey players

Variables	Elite players $n = 12$		County players $n = 12$		Whole sample $n = 24$	
Lung function						
VC (l – BTPS)	4.45[a] ±	0.41[b]	4.13 ±	0.49	4.29 ±	0.48
VC above norm (%)	17.0 ±	8.3	15.7 ±	8.6	16.4 ±	8.3
FEV_1 (l – BTPS)	3.81 ±	0.32	3.59 ±	0.33	3.70 ±	0.34
FEV_1 above norm (%)	13.6 ±	8.0	13.0 ±	6.5	13.5 ±	7.0
FEV_1 % VC	85.9 ±	5.1	87.5 ±	5.5	86.6 ±	5.3
Aerobic fitness						
Maximal heart rate (beats/min^{-1})	194.0 ±	6.8	190.0 ±	7.3	192.0 ±	7.1
PWC_{170} (kpm/min^{-1})	839.9 ±	108.8	739.0 ±	77.3	789.4 ±	105.7
PWC_{170} (kpm/kg^{-1}/min^{-1})	13.9 ±	2.3	12.4 ±	1.2	13.2 ±	1.9
$\dot{V}O_2max$ (l/min^{-1})	2.755 ±	0.464	2.408 ±	0.384	2.581 ±	0.453
$\dot{V}O_2max$ (ml/kg^{-1}/min^{-1})	45.7 ±	8.9	40.6 ±	6.1	43.2 ±	7.9
Field tests						
Sprint time (s)	6.99 ±	0.44	7.77 ±	0.65	7.38 ±	0.67
T-run time (s)	25.85 ±	1.94	29.04 ±	2.49	27.44 ±	2.72
Distance run (m)	360.6 ±	14.8	353.5 ±	17.3	357.1 ±	16.1
Accuracy (%)	78.89 ±	9.29	74.76 ±	6.02	76.83 ±	7.92

[a]Means; [b]standard deviations.

The only other significant relationship among components was for Accuracy and Linearity ($r = -.633$; $p < .001$).

Results of univariate F tests between the two levels of proficiency for each component showed the elite squad was superior in Dribbling Speed ($F = 13.62$; $p < .01$), Aerobic Fitness ($F = 5.38$; $p < .05$), and Anaerobic Power ($F = 6.64$; $p < .01$). The F value for examining the overall difference between the groups was nonsignificant. None of the components showed a significant F value for a difference among the three positional roles. Similarly, the result for an overall F was nonsignificant.

Discussion

The somatotype of subjects was close to that recently reported by other workers (Bale & McNaught-Davis, 1983), and it seems that a mean profile of approximately 3.5:4:2.5 is generally found in female hockey teams. Though the physique of subjects was not pronounced on any of the three dimensions, a marginal trend towards endomorphy was noted. A muscular physique would be beneficial in many aspects of the game such as tackling or two-handed hitting. A specific disadvantage of a linear physique was evident in the significant negative correlation of linearity and accuracy. Body fat was marginally lower than the 25.1% (Johnston & Watson, 1968) and 25.3% (Withers & Roberts, 1981) reported for national standard players. Field hockey players seem to have appreciably more body fat than female endurance runners of comparable caliber.

It seems that anthropometric factors lack the sensitivity to discriminate between elite and county standard players. It is likely, therefore, that players without the appropriate physical endowment might not attain intercounty representative status and would not progress beyond lower levels of club play. Nor could anthropometric variables distinguish between positional roles significantly. This suggests that modern hockey squads are relatively homogeneous in kinanthropometric make-up, which does not predispose individuals to specific playing positions.

The estimated VO_2max of subjects is compatible with the body composition data, although values were approximately 10% lower than those reported for American (Zeldis et al., 1978) and Australian (Withers & Roberts, 1981) international players. Mean PWC_{170} values were approximately 24% better in the elite squad than in sedentary females (Davies & Daggett, 1977) and were 13% superior to the values of the county players. It seems that aerobic fitness does exert an influence on the standard of play sustained by female hockey players, a tenet supported by the multivariate analyses.

The series of principal component analyses reduced the 30 fitness variables to 10 components. The component called Dribbling Speed involved an interplay with physical fitness measures because sprinting speed loaded highly in it. It discriminated between the levels of play, suggesting that ability to dribble a ball at speed is important. The combination of variables contributing to the component indicates that good sprinting ability is necessary to execute this skill. The importance of sprinting speed is evident also in the success of

Anaerobic Power in discriminating between squads, the two components being significantly correlated. The Anaerobic Power of the elite players on the stair test was similar to that of Australian internationals (Withers & Roberts, 1981), and greater than that of Indian national players (Verma et al., 1979), where a lower body mass may have contributed to poorer values. The benefit of high anaerobic power output is apparent when the frequent demands to accelerate, decelerate, and change direction in the game context are considered.

The success of Aerobic Fitness in separating the two levels of proficiency has repercussions for hockey training. Results implicate a need for regimes that incorporate aerobic as well as anaerobic routines. The superior aerobic fitness of the elite squad may have been in part due to its systematized training program, although the difference between groups was not reflected in body composition and flexibility. Irrespective of causes, the better fitness profile of the elite players confirms the importance of aerobic status for that level of play.

Failure of any fitness component to distinguish between the positional roles runs counter to previous reports (Johnston & Watson, 1968; Verma et al., 1979). The studies referred to were conducted before the revised rules of play promoted a style described as "total hockey." Differences in work rate between individuals may have become blurred with the changes introduced, and team members may now be more homogeneous in fitness status and more versatile than was conventionally the case. Though present data demonstrate clear differences in fitness level between two standards of proficiency, further study of a larger sample is advised to confirm the lack of specificity of positional role in kinanthropometry of female field hockey players.

References

Åstrand, P.O., & Ryhming, I. (1954). A nomogram for calculation of aerobic capacity (physical fitness) from pulse rate during submaximal work. *Journal of Applied Physiology,* **17**, 218.

Bale, P., & McNaught-Davis, P. (1983). The physiques, fitness and strength of top class hockey players. *Journal of Sports Medicine and Physical Fitness,* **23**, 80-88.

Child, D. (1970). *The essentials of factor analysis.* New York: Holt, Rinehart and Winston.

Davies, B., & Daggett, A. (1977). Responses of adult women to programmed exercise. *British Journal of Sports Medicine,* **11**, 122-126.

Durnin, J.V.G.A., & Womersley, J. (1974). Body fat assessed from total body density and its estimation from skinfold thickness measurements on 481 men and women aged 16-72 years. *British Journal of Nutrition,* **32**, 77-97.

Heath, B.H., & Carter, J.E.L. (1967). A modified somatotype method. *American Journal of Physical Anthropology,* **27**, 57-74.

Johnston, R.E., & Watson, J.M. (1968). A comparison of the phenotypes of women basketball and hockey players. *New Zealand Journal of Health, Physical Education and Recreation,* **40**, 48-54.

Maksud, M.G., Canninstra, C., & Dublinski, D. (1976). Energy expenditure and VO₂max of female athletes during treadmill exercise. *Research Quarterly,* **47**, 692-697.

Verma, S.K., Mohindroo, S.R., & Kansal, D.K. (1979). The maximal anaerobic power of different categories of players. *Journal of Sports Medicine and Physical Fitness,* **19**, 55-62.

Withers, R.T., & Roberts, R.G. (1981). Physiological profiles of representative women softball, hockey and netball players. *Ergonomics,* **24**, 583-591.

Zeldis, S.M., Morganroth, J., & Rubler, S. (1978). Cardiac hypertrophy in respect to dynamic conditioning in female athletes. *Journal of Applied Physiology: Respiratory Environmental and Exercise Physiology,* **44**, 849-852.

13

A Study of Intrasport Differences in the Physique of Indian University Football Players

Devinder K. Kansal
PUNJABI UNIVERSITY
PATIALA, INDIA

Neelam Gupta and Anil K. Gupta
GOVERNMENT RAJINDRA MEDICAL COLLEGE
PATIALA, INDIA

It is now an established fact that champions of different athletic and sporting events differ significantly in their physical and physiological characteristics that correspond, to some extent, with the particular requirements of their respective events (Borms & Hebbelinck, in press; Carter, Ross, Aubry, Hebbelinck, & Borms, 1982; de Garay, Levine, & Carter, 1974; Hirata, 1979; Sodhi & Sidhu, 1984; Tanner, 1964; Verma, Sidhu, & Kansal, 1979). Several investigators (Bouchard & Malina, 1983; Hirata, 1979; Kansal, 1982; Klissouras, 1971; Tanner, 1964) have suggested that physical training and other extragenetic factors can improve the performance of an athlete only up to a certain limit that is set by his genotype. Hence, it has become very important to select athletes with due consideration to their genetic performance potentialities. While doing so, the prerequisite selection criteria demand proper weighting of both the less flexible skeletal framework of the body and the relatively more adaptable physiological characteristics (Bouchard & Lortie, 1984; Klissouras, 1971).

It has also been demonstrated that it is not only the intersport differences which are significant, but also some of the body parameters, even within the same sport, depending upon playing position (Bell, 1973; Kansal, Verma, & Sidhu, 1980a, 1980b; Sodhi & Sidhu, 1984). It is thus important to discover the intrasport differences in physique and body composition which relate to differences in field positions within a team sport. This information would enable sports counselors and coaches to effectively advise talented candidates regarding the selection of appropriate field positions based on their specific latent potential. The purpose of this study was to determine the specific morphological characteristics of forward line, half line, back line, and goalkeeper football (soccer) players of Indian universities.

Method

The study was based on 151 football players of 12 Indian university teams including the champion teams of the four zones: North, East, South, and West India. The remaining eight teams were all from the North zone, as the data were collected during the intranorth zone competition and interzonal football competition of the Indian universities. The sample included 53 forwards, 34 half-liners, 47 stoppers/backs, and 17 goalkeepers. Though a total of 32 measurements were taken on each subject, only 16 measurements were considered in this study. These included weight, height, and sitting height; biacromial, bicristal, bicondylar humerus, and bicondylar femur diameters; upper arm, maximum chest, minimum chest, and calf circumferences; and triceps, midaxillary, suprailiac, subscapular, and calf skinfolds. All body measurements were taken following standard techniques (Weiner & Lourie, 1969). Bilaterally represented measurements were taken on the left side of the subject. Skinfold thicknesses were measured using a Harpenden skinfold caliper. The physique of each subject was evaluated by the Heath-Carter method (1967). In addition, the somatotype ratings of each subject were computed applying the recently suggested modifications to the above method (Kansal, 1983, 1985).[1]

Results

The means and standard deviations of the 16 variables studied in four intrasport categories of footballers based on their field positions are shown in Table 1 along with the corresponding values of the total sample. It may be observed from this table that, generally speaking, the forwards and half-liners have

---

[1]Repetition here of the rationale for these modifications is redundant, but it is pertinent to mention briefly the modified procedures. For assessing mesomorphy, the value of $\pi \times$ skinfold thickness is used in the correction of upper arm girth and calf girth. For assessing ectomorphy the Corrected Ponderal Index is used. Corrected Ponderal Index = (Height/ $^3\sqrt{\text{Weight}}$) − (10.65 − Height/BHB + BFB) where BHB = bicondylar humerus breadth, and BFB = bicondylar femur breadth. Note that the constant 10.65 is the value of Height/BHB + BFB for the unisex universal Phantom reported by Ross and Wilson (1974).

Table 1. Means and standard deviations of anthropometric measures for forward line (F), half line (H), back line (B), and goalkeeper (G) football players of Indian universities

Variable n =	Mean					Standard deviation				
	F 53	H 34	B 47	G 17	Total 151	F 53	H 34	B 47	G 17	Total 151
Weight (kg)	52.9	53.1	59.0	56.0	55.2	5.7	6.6	5.5	4.8	6.3
Height (mm)	1659	1675	1716	1735	1689	43	60	51	41	57
Sitting height (mm)	864	873	897	890	877	30	34	49	26	39
Diameters (mm)										
Biacromial	372	371	385	379	377	22	18	20	12	20
Bicristal	254	255	263	263	258	16	16	11	11	17
Bicondylar humerus	65.0	65.0	67.6	67.8	66.1	2.6	3.3	2.6	3.5	3.0
Bicondylar femur	91.0	91.6	94.1	93.3	92.4	5.7	4.4	3.7	2.6	4.3
Circumferences (mm)										
Upper arm	241	237	251	244	243	17	13	16	11	16
Chest maximum	857	858	892	880	871	38	38	41	17	40
Chest minimum	821	820	846	830	830	37	41	36	28	43
Calf	323	321	338	346	328	11	20	27	18	21
Skinfolds (mm)										
Triceps	5.1	5.1	5.6	4.9	5.2	1.0	1.4	1.3	0.8	1.3
Midaxillary	4.8	4.9	5.1	4.5	4.9	0.9	1.4	1.0	0.6	0.6
Suprailiac	6.0	6.5	6.3	5.3	6.1	2.4	1.3	1.9	0.8	2.7
Subscapular	7.3	7.5	7.8	6.9	7.4	1.6	1.4	1.8	0.8	1.7
Calf	4.8	4.9	5.2	4.7	4.9	1.1	1.6	1.3	0.9	1.3
Somatotype ratings										
Endomorphy	1.99	2.03	2.05	1.81	1.91	0.6	0.8	0.5	0.4	0.9
Mesomorphy										
(a) Heath-Carter	3.67	3.44	3.63	3.35	3.57	0.5	0.5	0.6	0.5	0.5
(b) Modified	3.43	3.22	3.43	3.07	3.35	0.5	0.6	0.6	0.6	0.4
Ectomorphy										
(a) Heath-Carter	4.06	4.35	3.96	4.72	4.17	1.2	0.9	1.1	0.9	1.1
(b) Modified	4.02	4.37	3.95	4.82	4.17	1.2	1.1	0.8	1.1	1.1

comparable values for most of the size measures, especially the breadths and circumferences; back-liners (stoppers) have somewhat closer values to those of the goalkeepers except for skinfolds, where the former usually possess the maximum and the latter usually have the minimum values among the four subgroups of football players. Thus, forwards are lighter and shorter by just 0.2 kg and 16 mm, respectively, than the half-liners, but by 6.1 kg and 57 mm than the backs, and by 3.1 kg and 76 mm, respectively, than the goal-keepers.

However, the picture changes sharply in the case of relative body shape as reflected by the somatotype ratings: Forwards have values closer to those of the backs, and half-liners are more similar to goalkeepers (Table 1). Somatotype status of backs and goalkeepers is quite different, unlike breadth and cir-cumference measures where they are the most similar. The backs' average rating is 2.05:3.63:3.96 whereas the goalkeepers' mean physique rating is found to be 1.81:3.35:4.72; the backs have larger endomorphy and mesomorphy components but are less ectomorphic than the goalkeepers. It is also evident from Table 1 that the average values of the total sample are quite different from those of the individual categories of players.

The mean differences and the values of Student t-tests have been listed in Table 2 to provide a comprehensive view of the differences among the four subgroups of football players and to evaluate the statistical significance of these within-sport differences. The following is evident from Table 2:

1. Forwards and half-liners have significantly lower values than those of backs and goalkeepers for most of the anthropometric measurements studied ex-cept the skinfolds. However, forwards do not differ from the backs, and half-liners do not differ either from backs or from goalkeepers in any of the somatotype components.
2. Backs and goalkeepers differ significantly only in body weight and in four of the five skinfolds studied. However, in none of the other anthropometric measurements studied, including heights, diameters, and circumferences, are the differences significant.
3. No differences between forward and half-line players are significant ex-cept in the case of mesomorphy, where the forwards are significantly more mesomorphic than the half-line players when studied using the Heath-Carter method of somatotyping, but not when the rating is based on Kansal's (1983) modification.
4. Generally speaking, body shape has shown nonsignificant differences among the four subgroups of footballers with the exceptions that forwards and backs are found to be significantly more mesomorphic and less ectomorphic than the goalkeepers.

The physical status of each intrasport category of footballers was also com-pared with the average status of the total sample of footballers; the values of mean differences and t-scores thus obtained are shown in Table 3. It is in-teresting to note that the backs and goalkeepers have generally higher mean values than the mean values of the total sample and that many of these differences are statistically significant.

Table 2. Mean differences between groups of forwards (F), half-liners (H), backs (B), and goalkeepers (G) compared on anthropometric measures and somatotypes

Variable	F vs H		F vs B		F vs G		H vs B		H vs G		B vs G	
	MD	t	MD	t	MD	t	MD	t	MD	t	MD	t
Weight (kg)	-0.2	0.12	-6.1	5.42*	-3.1	2.22*	-5.9	4.29*	-2.9	1.83	3.0	2.13*
Height (mm)	-16	1.31	-57	5.92*	-76	6.54*	-41	3.23*	-60	4.23*	-19	1.57
Sitting height (mm)	-9	1.20	-33	4.01*	-26	3.47*	-24	1.94	-17	2.02*	7	0.04
Diameters (mm)												
Biacromial	1	0.26	-13	3.14*	-7	1.73	-14	3.35*	-8	1.95	6	1.38
Bicristal	-1	0.40	-9	3.30*	-9	2.92*	-8	2.41*	-8	2.20*	0	0
Bicondylar humerus	0.0	0.00	-2.6	4.91*	-2.8	3.04*	-2.6	2.35*	-2.8	2.77*	-0.2	0.22
Bicondylar femur	0.6	0.56	-3.1	3.30*	-2.3	2.32*	-2.5	2.69*	-1.7	1.74	0.8	0.98
Circumferences (mm)												
Upper arm	4	1.04	-10	3.13*	-3	1.04	-14	4.20*	-7	2.02*	7	1.90
Chest maximum	-1	0.20	-35	4.43*	-23	3.42*	-34	3.78*	-22	2.73*	12	1.86
Chest minimum	1	0.12	-25	3.47*	-9	1.02	-26	3.00*	-10	0.98	16	1.91
Calf	2	0.62	-15	2.37*	-23	4.83*	-17	4.19*	-25	4.45*	-8	1.23

(Cont.)

Table 2. (Cont.)

	MD		MD		MD		MD		MD		MD	
Skinfolds (mm)												
Triceps	0.0	0.00	-0.5	1.80	0.2	1.12	-0.5	1.47	0.2	0.48	0.7	2.59*
Midaxillary	-0.1	0.14	-0.3	1.62	0.3	1.60	-0.2	0.72	0.4	0.98	0.6	3.08*
Suprailiac	-0.5	1.73	-0.3	0.53	0.7	1.91	0.2	0.87	1.2	3.14*	1.0	2.85*
Subscapular	-0.2	0.87	-0.5	1.59	0.4	1.39	-0.3	0.96	0.6	1.64	0.9	2.90*
Calf	-0.1	0.22	-0.4	1.71	0.1	0.15	-0.3	1.01	0.2	0.31	0.5	1.61
Somatotype ratings												
Endomorphy	-0.04	0.18	-0.06	0.49	0.18	1.32	-0.02	0.10	0.22	1.79	0.24	1.85
Mesomorphy												
(a) Heath-Carter	0.23	2.02*	0.04	0.14	0.32	2.39*	-0.19	1.53	-0.09	0.46	0.28	1.95
(b) Modified	0.21	1.75	0.00	0.00	0.36	2.36*	-0.21	1.61	0.15	0.91	0.36	2.23*
Ectomorphy												
(a) Heath-Carter	-0.29	1.25	0.10	0.46	-0.66	2.35*	0.39	1.73	-0.37	0.24	-0.76	2.75*
(b) Modified	-0.35	1.42	0.07	0.33	-0.80	2.55*	0.42	1.88	-0.45	1.34	-0.87	2.94*

Note. MD = mean difference.
*p < .05.

Table 3. Mean differences between groups of forwards (F), half-liners (H), backs (B), and goalkeepers (G) compared separately with the total sample of Indian university football players on anthropometric measures and somatotype

Variable	F MD	F t	H MD	H t	B MD	B t	G MD	G t
Weight (kg)	-2.3	2.43*	-2.1	1.71	3.8	3.99*	0.8	0.65
Height (mm)	-30	3.94*	-14	1.26	27	3.05*	46	4.21*
Sitting height (mm)	-13	2.54*	-4	0.67	20	2.52*	13	2.55*
Diameters (mm)								
Biacromial	-5	1.35	-6	1.62	8	2.59*	2	0.57
Bicristal	-4	1.60	-3	0.91	5	2.28*	5	1.80
Bicondylar humerus	-1.1	2.56*	-1.1	1.80	1.5	3.31*	1.7	1.93
Bicondylar femur	-1.4	1.65	-0.8	0.96	1.7	2.64*	0.9	1.48
Circumferences (mm)								
Upper arm	-2	1.08	-0.6	2.38*	8	2.78*	1	0.25
Chest maximum	-9	1.71	-13	1.66	21	3.14*	9	1.71
Chest minimum	-14	1.35	-10	1.51	16	2.58*	0	0.00
Calf	-5	1.92	-7	1.76	10	2.32*	18	3.83*
Skinfolds (mm)								
Triceps	-0.1	0.80	-0.1	0.61	0.4	1.75	-0.3	1.38
Midaxillary	-0.1	0.98	0.0	0.00	0.2	1.51	-0.4	2.34*
Suprailiac	-0.1	0.25	0.4	1.27	0.2	0.58	-0.8	2.86*
Subscapular	-0.1	0.63	0.1	0.18	0.4	1.27	-0.5	2.32*
Calf	-0.1	0.76	0.0	0.00	0.3	1.28	-0.2	0.82
Somatotype ratings								
Endomorphy	0.08	0.73	0.12	1.03	0.14	1.33	-0.10	0.80
Mesomorphy								
(a) Heath-Carter	0.10	1.25	-0.13	1.28	0.06	0.63	-0.22	1.76
(b) Modified	0.08	0.98	-0.13	1.29	0.08	0.83	-0.28	2.02*
Ectomorphy								
(a) Heath-Carter	-0.11	0.59	0.18	0.97	-0.21	1.17	0.55	2.26*
(b) Modified	-0.15	0.81	0.20	0.96	-0.22	1.48	0.65	2.87*

*$p < .05$.

Discussion

The findings of the present investigation are in general agreement with those of the previous studies carried out by Bell (1973), Verma et al. (1979), Kansal et al. (1980a, 1980b), and Sodhi and Sidhu (1984), all of which indicate that players performing different duties in a team sport may differ considerably in their physical status. These results thus demand reconsideration of the common standards used for all the players on a team. Based on the average height of 263 Montreal Olympic footballers, Hirata (1979), for example, has recommended a height of 175.5 cm or above for the selection of candidates for Olympic football. The authors, however, are of the view that such a recommendation does not give due consideration to the wide range in stature of the Montreal Olympic footballers, ranging from 155.0 cm to 192.5 cm.

In light of the findings of significant intrasport differences in the present study, the wide distribution of height among the players of a single sport may be partially due to the differential positional requirements of that sport. The forward and half-liners might be concentrated quite below the observed average of 175.5 cm, while backs and goalkeepers might be mainly scattered above this average. The above hypothesis appears to hold true when the ranges of variables for the intrasport groups of footballers are studied in relation to the ranges of the total sample (see Table 4). The ranges in the total sample are much wider than those observed in the individual categories. For instance, the height of goalkeepers ranges only from 1,667 mm to 1,825 mm (158 mm) and that of backs, forward, and half-line players from 1,607 mm to 1,803 mm, 1,555 mm to 1,752 mm, and 1,559 mm to 1,813 mm, respectively. In comparison, the range of height for the total sample is from 1,555 mm to 1,825 mm (270 mm). Similarly, stature ranges from 155.0 cm to 192.5 cm in the Montreal Olympic footballers (Hirata, 1979) and even from 153.4 cm to 202.5 cm in the male athletes of the MOGAP sample (Carter et al., 1982).

The observation of such a wide range among individuals usually described as possessing specific physical characteristics may be partly due to significant intrasport as well as intersport differences. The situation thus demands that researchers develop differential standards for different categories of athletes with due consideration to both intersport and intrasport differences. There are numerous studies undertaken on different categories of athletes, demonstrating that athletes of various categories have quite specific physical status, which is different from that of nonathletes (Borms & Hebbelinck, 1984; Carter et al., 1982; de Garay et al., 1974; Hirata, 1979; Tanner, 1964). Although the above studies also include large scale comprehensive reports of physical measures of Olympic athletes, they are still insufficient to provide adequate standards, especially in the case of team sports.

Because data on Olympic athletes are rare and difficult to obtain, researchers have invariably pooled all available data. For instance, Carter et al. (1982) reported the mean values of 30 anthropometric variables of 309 male and 148 female Olympic athletes, the former distributed over 78 events and the latter over 32 events. The averages reported were quite skewed in accordance with the composition of the sample studied. The reported average height of 180.3 cm, for example, was the result of large numbers of rowers included in the

Table 4. Ranges of various anthropometric measurements and somatotype ratings in forward line (F), half line (H), back line (B), and goalkeeper (G) football players of Indian universities

Variable		Range				
N =		F 53	H 34	B 47	G 17	Total 151
Weight (kg)		41.0 - 67.0	40.5 - 73.0	49.5 - 68.0	48.0 - 67.5	40.5 - 73.0
Height (mm)		1555 - 1752	1559 - 1813	1607 - 1803	1667 - 1825	1555 - 1825
Sitting height (mm)		770 - 928	797 - 980	836 - 964	827 - 921	770 - 980
Diameters (mm)						
Biacromial		328 - 467	330 - 400	350 - 419	362 - 404	328 - 467
Bicristal		207 - 288	232 - 280	250 - 295	235 - 278	207 - 295
Bicondylar humerus		59 - 70	59 - 70	62 - 72	61 - 76	59 - 76
Bicondylar femur		71 - 103	82 - 100	86 - 101	90 - 97	71 - 103
Circumferences (mm)						
Upper arm		204 - 284	214 - 265	222 - 281	226 - 268	204 - 284
Chest maximum		808 - 965	790 - 955	783 - 990	855 - 911	783 - 990
Chest minimum		725 - 916	740 - 920	734 - 935	790 - 892	725 - 935
Calf		293 - 378	272 - 355	290 - 372	307 - 356	272 - 378
Skinfolds (mm)						
Triceps		3.0 - 8.0	3.8 - 11.2	4.0 - 9.6	4.2 - 7.2	3.0 - 11.2
Midaxillary		3.3 - 7.0	3.6 - 12.0	4.0 - 6.8	3.4 - 5.6	3.3 - 12.0
Suprailiac		3.4 - 17.8	3.8 - 11.8	4.1 - 11.2	4.4 - 6.4	3.4 - 11.8
Subscapular		4.4 - 11.6	4.8 - 10.6	5.5 - 13.4	5.8 - 8.2	4.4 - 13.4
Calf		3.0 - 6.8	3.4 - 11.4	3.6 - 9.6	3.6 - 7.0	3.0 - 11.4
Somatotype ratings						
Endomorphy		1.1 - 3.7	1.3 - 6.1	1.2 - 5.6	1.4 - 3.2	1.1 - 6.1
Mesomorphy						
(a) Heath - Carter		2.4 - 5.2	2.4 - 4.1	2.2 - 4.7	2.4 - 4.0	2.2 - 5.2
(b) Modified		2.2 - 4.9	2.4 - 4.0	2.0 - 4.5	1.8 - 4.0	1.8 - 4.9
Ectomorphy						
(a) Heath - Carter		2.1 - 6.2	2.5 - 6.4	2.3 - 6.7	2.1 - 6.1	2.1 - 6.7
(b) Modified		1.8 - 6.3	2.2 - 6.9	1.9 - 6.4	2.0 - 6.6	1.8 - 6.9

sample. Had there been more weightlifters and gymnasts, the mean stature would undoubtedly have been much lower than that reported. With this in mind, it is suggested that overall average values of Olympic or any other level athletes should be described with due care, especially when making comparisons with other samples of similar studies.

Conclusions

Two conclusions were derived from this study.

1. The reporting of overall average physical structure of champion athletes without giving due consideration to the composition of the sample may be of little value.
2. Immediate and careful attention of sport scientists is needed to establish differential standards for the physical structure of various categories of players, performing somewhat different types of activities in the same field of play. This requires the examination of large numbers of athletes belonging to national and international championship teams.

References

Bell, W. (1973). Distribution of skinfolds and differences in body proportions in young adult rugby players. *The Journal of Sports Medicine and Physical Fitness, 13*, 69-73.

Borms, J., & Hebbelinck, M. (1984). Review of studies on Olympic athletes. In J.E.L. Carter (Ed.), *Physical structure of Olympic athletes: Part II. Kinanthropometry of Olympic athletes*. Switzerland: S. Karger.

Bouchard, C., & Malina, R.M. (1983). Genetics of physiological fitness and motor performance. *Exercise & Sports Sciences Reviews, 11*, 306-339.

Bouchard, C., & Lortie, G. (1984). Heredity and endurance performance. *Sports Medicine, 1*, 38-64.

Carter, J.E.L., Ross, W.D., Aubry, S.P., Hebbelinck, M., & Borms, J. (1982). Anthropometry of Montreal Olympic athletes. In J.E.L. Carter (Ed.), *Physical structure of Olympic athletes: Part I. The Montreal Games Anthropological Project* (pp. 25-52). Switzerland: S. Karger.

de Garay, A.L., Levine, L., & Carter, J.E.L. (1974). *Genetic and anthropological studies of Olympic athletes*. New York: Academic Press.

Heath, B.H., & Carter, J.E.L. (1967). A modified somatotype method. *American Journal of Physical Anthropology, 27*, 57-74.

Hirata, K.I. (1979). *Selection of Olympic champions* (Vols. 1 & 2). Toyota: Chukyo University Press.

Kansal, D.K. (1982). A scientific way for the prediction of adult physical status of male children and its importance to sportsmen [in Punjabi]. *Vigyan De Nakas, 16*, 30-38.

Kansal, D.K. (1983, January). Modifications to Heath-Carter method of somatotyping. Paper presented in 70th session of Indian Science Congress Association (Abstract No. 56), Tirupati.

Kansal, D.K. (1985). Growth patterns of physique in Jat-Sikh and Bania males of Punjab [India] and some modifications to Heath-Carter method of somatotyping. Under preparation.

Kansal, D.K., Verma, S.K., & Sidhu, L.S. (1980a). Anthropometric characteristics of Indian university zonal champion football players. *The Journal of Sports Medicine and Physical Fitness,* **20**, 275-284.

Kansal, D.K., Verma, S.K., & Sidhu, L.S. (1980b). Intrasport differences in maximum oxygen uptake and body composition of Indian players in hockey and football. *The Journal of Sports Medicine and Physical Fitness,* **20**, 309-316.

Klissouras, V. (1971). Heritability of adaptive variation. *Journal of Applied Physiology,* **31**, 338-344.

Ross, W.D., & Wilson, N.C. (1974). A stratagem for proportional growth assessment. *Acta Paediatrica Belgica.,* **28** (Suppl.), 169-182.
Paediatrica Belgica., **28** (Suppl.), 169-182.

Sodhi, H.S., & Sidhu, L.S. (1984). *Physique and selection of sportsmen.* Patiala: Punjab Publishing House.

Tanner, J.M. (1964). *The physique of the Olympic athlete.* London: Allen & Unwin.

Verma, S.K., Sidhu, L.S., & Kansal, D.K. (1979). The maximum anaerobic power of different categories of players. *The Journal of Sports Medicine and Physical Fitness,* **19**, 55-62.

Weiner, J.S., & Lourie, J.A. (Eds.). (1969). *Human biology: A guide to field methods.* Oxford: Blackwell Scientific Publications.

14

Body Structure, Somatotype, and Motor Fitness of Top-Class Belgian Judoists

Albrecht L.M. Claessens, Gaston P. Beunen,
Jan M. Simons, Rita I. Wellens,
Dirk Geldof, and Marina M. Nuyts
KATHOLIEKE UNIVERSITEIT LEUVEN
LEUVEN-HEVERLEE, BELGIUM

The relation between body structure and function has already been stressed in a number of investigations. Apart from other factors (e.g., technical-tactical, psychological, physiological), the bodily constitution also plays a determining role in the achievement of top sporting performances. Studies on judoists (Azuma, 1964; Carter, 1982; Carter, Aubry, & Sleet, 1982a; Carter, Ross, Aubry, Hebbelinck, & Borms, 1982b; Chernilo, Soto, & Fernandez, 1979; Claessens, Beunen, Wellens, & Geldof, 1983; Farmosi, 1980; Hirata, 1966, 1979; Maas, 1974; Matsumoto et al., 1969; Mery & Marbach, 1979; Ross, Ward, Leahy, & Day, 1982; Taylor & Brassard, 1981; Toteva & Slanchev, 1983) show that an outstanding judoist can be characterized as a robustly built athlete with a rather "thick-set" stature, with fairly large breadth and girth measurements, and with little development of subcutaneous fat. In general, the body build of a top judoist can be situated in the endomesomorph zone on the somatochart. However, little attention has been given to the motor performance of judoists (Farmosi, 1980; Matsumoto et al., 1969; Taylor & Brassard, 1981). The aim of this study was to describe the somatic characteristics, the somatotype, and motor performance of top-class Belgian judoists, taken as a whole group and per weight category.

Method

The test group consisted of 24 top-class Belgian judoists varying from 17.0 to 29.3 years of age with a mean age of 21.9 years. For this investigation only, the total group was divided into two weight categories: under 71 kg (n = 13) and 71 to 86 kg (n = 9). Because only two subjects were above 86 kg, this category was not considered separately in this study.

A large number (23) of anthropometric measurements was taken to determine the somatic structure of the judoists. The measurements were taken by two experienced kinanthropometrists according to the instructions described by Cameron (1978) and Renson, Beunen, van Gerven, Simons, and Ostyn (1980). Apart from the somatic variables, two bodily indices were calculated: the masculinity index (biiliac diameter × 100/biacromial diameter) and the ponderal index (height/$\sqrt[3]{}$ weight). The three components of the somatotype were photoscopically determined according to the Leuven method which is a modified "Atlas of Men" (Sheldon, Dupertuis, & McDermott, 1954) technique. (For a detailed description of this method, see Claessens et al., 1980.) To evaluate motor fitness, a motor ability test battery was administered to the subjects. A detailed description of the tests is given by Renson et al. (1980).

Results

Somatic Structure

Table 1 gives the means and standard deviations of the anthropometric characteristics, both for the group of judoists as a whole and per weight category. The somatic profiles of the whole group and for the two weight classes separately, plotted against Belgian reference data for 19- to 19.5-year-old boys (Ostyn, Simons, Beunen, Renson, & van Gerven, 1980), are given in Figure 1. Compared to Belgian reference data of young adults, judoists are heavy

Table 1. Means (M) and standard deviations (SD) of somatic characteristics of Belgian judoists

Variable	Total sample (N = 24) M	SD	Group < 71 kg (n = 13) M	SD	Group 71 to 86 kg (n = 9) M	SD
Weight (kg)**	74.3	10.9	65.9	4.8	81.2	3.4
Height (cm)**	175.2	7.2	170.7	4.3	178.1	4.0
Sitting height (cm)*	91.6	4.3	88.9	2.7	93.8	2.9
Leg length (cm)	83.6	4.0	81.8	3.1	83.3	3.2
Forearm length (cm)	25.1	1.3	24.6	1.0	25.5	1.4
Head length (cm)	19.2	0.6	18.9	0.6	19.3	0.4
Head breadth (cm)	15.2	0.4	15.1	0.4	15.3	0.5
Head circumf. (cm)	56.5	1.4	55.9	1.3	57.2	0.9
Biacrom. diam. (cm)	39.1	2.3	37.9	1.5	40.1	1.7
Chest width (cm)	31.7	2.3	30.2	1.9	33.4	1.7
Biiliac diam. (cm)	27.6	1.8	26.2	0.9	28.9	0.9
Hum. width (cm)	7.3	0.3	7.0	0.2	7.4	0.3

(Cont.)

Table 1. (Cont.)

Femur width (cm)	9.8	0.5	9.5	0.3	10.3	0.2
Triceps skinf. (mm)	6.2	1.8	6.2	2.0	6.2	1.6
Subscap. skinf. (mm)	8.2	2.0	7.9	1.6	8.3	1.6
Suprailiac skinf. (mm)	5.2	1.2	5.5	1.3	5.4	1.0
Biceps skinf. (mm)	3.9	0.7	3.8	0.8	4.0	0.6
Calf skinf. (mm)	6.7	2.1	6.2	1.5	7.1	2.9
Sum skinfolds[a] (mm)	19.4	5.0	19.3	4.7	18.9	5.1
Flex. arm girth (cm)	33.5	2.5	31.8	1.5	35.2	1.9
Ext. arm girth (cm)	30.2	2.5	28.5	1.5	32.1	1.9
Forearm girth (cm)	28.0	1.8	26.8	1.2	29.0	0.9
Thigh girth (cm)	55.6	3.6	53.3	2.8	57.6	1.6
Calf girth (cm)	36.8	2.6	35.0	1.9	38.5	1.0
Masculinity index	70.7	3.6	69.5	3.6	72.2	3.0
Ponderal index	42.2	2.1	42.3	1.2	42.1	3.2
Endomorphy	—		3.0	0.8	3.1	0.6
Mesomorphy	—		4.8	0.6	5.6	0.5
Ectomorphy	—		2.4	0.9	1.7	0.6

Note. Significant difference between weight categories (* = $p < .05$; ** = $p < .01$); [a]sum skinfolds = triceps + subscapular + suprailiac skinfolds.

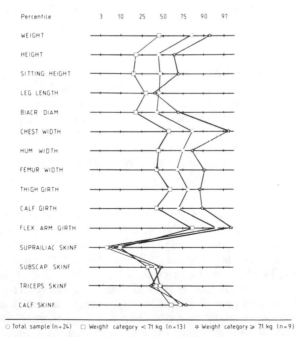

SOMATIC CHARACTERISTICS

○ Total sample (n=24)   □ Weight category < 71 kg (n=13)   ☆ Weight category ⩾ 71 kg (n=9)

Figure 1. Somatic profile of Belgian judoists for the total sample and per weight category.

for their stature and have average (about the 50th percentile) length measurements. Except for the biacromial diameter, all the mean values of their breadth measurements are situated near or above the 75th percentile of the reference data. Furthermore, these judoists have a large mean value for flexed upper arm girth; their mean value is situated between P_{90} and P_{97} of the reference data. Relatively smaller skinfold values are noticeable, especially the very low values for the suprailiac skinfold, of which the mean value is situated between the percentile values P_3 and P_{10}.

When we compare the somatic structures of the two weight categories with one another, the results indicate that except for the skinfolds, a rise in the mean values is found as the weight class increases; however, only for weight, height, and sitting height are differences significant. From Figure 1 it appears that within each category an almost identical profile of average body structure is characterized.

Proportionality Analysis

On the average these judoists show a similar masculinity index (biiliac width/biacromial width) to that of the average 20-year-old Belgian male, respectively 70.7 and 70.2. Although the higher weight category appeared to have a higher index value, no significant difference between the means was found. In comparison with 20-year-old Belgian males, Belgian judoists generally have a lower value for the ponderal index (43.2 and 42.2, respectively). Although no significant difference between the means of the two weight categories was demonstrated, there is a trend toward a lower mean ponderal index with greater mean weight.

Somatotype

In Figure 2 the individual somatoplots per weight category are arranged on a somatochart according to the method of Stephens and Taylor (1962). The somatotypes of judoists are principally localized in the endomesomorph area (classification according to Carter, 1975). Seventeen judoists are estimated as endomesomorph, 3 as midtypes, 3 as ectomesomorph, and 1 as mesoectomorph (see Table 1 for component values). It is also demonstrated that when the weight class increases, there is a shift from east to west on the somatochart.

Motor Performance

Means and standard deviations of the motor tests for the whole group and per weight category are given in Table 2. In comparison with a male Belgian reference population (Figure 3), the judoists achieved better results on the motor performance tests except for speed of limb movement (number of plate tappings in 20 s) in which the mean value is on the P_{50} value of the reference data. For the factors of flexibility (sit and reach), explosive strength (vertical jump), static strength (arm pull), and functional strength (bent arm hang), the mean values are around the P_{75} value. For the factor running speed (shuttle run), and especially for the factor trunk strength (leg lifts), judoists perform better in comparison with reference subjects, with mean values above P_{90} and P_{97}, respectively. For 16 judoists the leg lift results, for example, are better

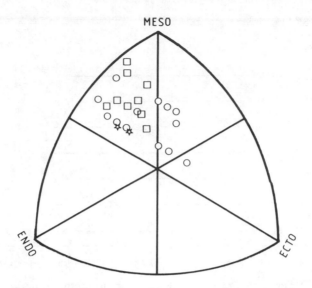

Figure 2. Individual somatoplots of Belgian judoists per weight category.
(● = < 71 kg; □ = 71 to 86 kg; ★ = +86 kg)

Table 2. Means (*M*) and standard deviations (*SD*) of motor characteristics of Belgian judoists

Motor test item	Total sample (N = 24) M	SD	Group < 71 kg (n = 13) M	SD	Group 71 to 86 kg (n = 9) M	SD
Plate tapping* (*n* in 20 s) (speed of limb movement)	92.9	10.0	90.5	9.3	96.3	11.0
Sit and reach* (cm) (flexibility)	30.7	7.1	28.0	6.3	34.2	7.6
Vertical jump (cm) (explosive strength)	52.5	6.7	53.3	6.4	50.2	7.4
Arm pull** (kg) (static strength)	84.3	13.4	77.2	11.3	90.4	10.6
Leg lifts (*n* in 20 s) (trunk strength)	21.3	1.7	21.9	1.3	21.1	1.7
Bent arm hang (sec) (functional strength)	48.4	11.9	50.7	10.2	47.6	14.3
Shuttle run 50 m (sec) (running speed)	19.0	0.6	18.9	0.7	19.2	0.5
Handgrip right (kg) (static strength)	64.9	8.9	56.8	7.7	59.7	6.1
Handgrip left* (kg) (static strength)	59.7	8.8	54.4	7.5	59.3	7.6

Note. Significant difference between weight categories (* = *p* < .05; ** = *p* < .01).

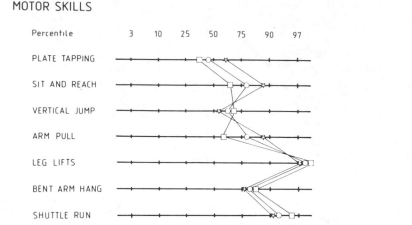

○ Total sample (n = 24)   □ Weight category < 71 kg (n = 13)   ✷ Weight category ⩾ 71 kg (n = 9)

Figure 3. Motor profile of Belgian judoists for the total group and per weight category.

than the P_{97} value, and for 5 others above the value of P_{90}; no judoist performed lower than the P_{75} value on this test. As seen in Figure 3 and Table 2, motor performance is not clearly related to the weight category, except for speed of limb movement (plate tapping), flexibility (sit and reach), and the static strength tests (arm pull and handgrip left). For these tests, significantly better results were obtained for the weight category 71 to 86 kg in comparison with the less than 71 kg weight class.

Discussion

As indicated by the somatic analysis, a top-class Belgian judoist can be characterized as a robustly built athlete, with large breadths and circumferential measurements in relation to his length, and with little subcutaneous fat. Almost an identical somatic profile emerges within each weight category. The findings of Matsumoto et al. (1969) on the Japanese national team and of Claessens et al. (1983) on 38 world-class judoists studied during the 1981 Senior World Championships at Maastricht in The Netherlands agree very well with findings of this study.

From the proportional analysis it is demonstrated that a Belgian judoist shows a similar masculinity index to that of a reference population, with a trend toward a higher index value among heavier judoists. Similar results were found by Claessens et al. (1983) on a group of 38 outstanding judoists, the mean index value varying from 69.8 for the lowest weight category to 73.5 for the highest weight category. Based on the mean values for the shoulder and pelvis breadths reported by Maas (1974) and Carter et al. (1982b), mean index values of 65.5 and 69.1, respectively, are obtained. These means are even lower than the

Table 3. Mean ponderal index values of judoists from selected, related studies

Study	Total sample	Weight category	
		High	Low
Chernilo et al. (1979)	—	41.3	39.0
(N = 49)		(n = 8)	(n = 3)
Claessens et al. (1983)	40.9	41.8	39.3
(N = 38)		(n = 18)	(n = 11)
Farmosi (1980)	40.3	41.4	39.8
(N = 18)		(n = 7)	(n = 11)
Hirata (1979)	—	42.0	38.9
(N = 167)		(n = 31)	(n = 25)

mean value of 69.5 for our judoists from the lowest weight category. That a judoist should be characterized by a "broad-based" trunk is not directly confirmed either by our findings or by those from the literature. Based on the results of the ponderal index, it is clearly demonstrated that a Belgian judoist is rather thick-set, and this characteristic is more pronounced within the heavier weight class. This trend was also observed by Chernilo et al. (1979), Claessens et al. (1983), Farmosi (1980), and Hirata (1979), as shown in Table 3.

The somatotype analysis shows that the typical Belgian judoist can be characterized as an endomesomorph, and that there is a shift from a more balanced mesomorph type for the lowest weight category to a more "heavy" endomesomorph type for the highest weight category. These findings parallel those of Carter et al. (1982a), Chernilo et al. (1979), Farmosi (1980), and Toteva and Slanchev (1983), in which the body type is anthropometrically determined following the Heath-Carter method, and by the study of Claessens et al. (1983) on world-class judoists in which the Leuven somatotype method was used.

Similarly, Belgian judoists also had better results on motor performance tests in comparison with a reference population, which was in accord with the findings of Taylor and Brassard (1981) on 19 Canadian judoists selected to attend the training camp for the 1979 Pan-American Games and of Farmosi (1980) on 18 judoists of the Hungarian national team. The results also indicate that motor performance is not directly related to the weight category, except for static strength tests (arm pull and handgrip left), speed of limb movement (plate tapping), and flexibility (sit and reach).

References

Azuma, T. (1964). *Olympic medical archives. Report Tokyo 1964.* Tokyo: International Olympic Medical Archives Committee.

Cameron, N. (1978). The methods of auxological anthropometry. In F. Falkner & J.M. Tanner (Eds.), *Human growth: Vol. II. Postnatal growth* (pp. 35-90). New York: Plenum.

Carter, J.E.L. (1975). *The Heath-Carter somatotype method*. San Diego, CA: San Diego State University.

Carter, J.E.L. (1982). Body composition of Montreal Olympic athletes. In J.E.L. Carter (Ed.), *Physical structure of Olympic athletes: Part I. The Montreal Olympic Games anthropological project: Vol. 16. Medicine and sport* (pp. 107-116). Basel: Karger.

Carter, J.E.L., Aubry, S.P., & Sleet, D.A. (1982a). Somatotypes of Montreal Olympic athletes. In J.E.L. Carter (Ed.), *Physical structure of Olympic athletes: Part I. The Montreal Olympic Games anthropological project: Vol. 16. Medicine and sport* (pp. 53-80). Basel: Karger.

Carter, J.E.L., Ross, W.D., Aubry, S.P., Hebbelinck, M., & Borms, J. (1982b). Anthropometry of Montreal Olympic athletes. In J.E.L. Carter (Ed.), *Physical structure of Olympic athletes: Part I. The Montreal Olympic Games anthropological project: Vol. 16. Medicine and sport* (pp. 25-52). Basel: Karger.

Chernilo, B., Soto, I., & Fernandez, A.A. (1979). *Composición corporal y somatotipo en judokas. Juegos Pan-americanos Puerto Rico 1979*. Unidad Médica Coch: Comité Olimpico de Chili.

Claessens, A., Beunen, G., Simons, J., Swalus, P., Ostyn, M., Renson, R., & van Gerven, D. (1980). A modification of Sheldon's anthroposcopic somatotype technique. *Anthropologiai Közlemények*, **24**, 45-54.

Claessens, A., Beunen, G., Wellens, R., & Geldof, D. (1983, August). Somatotype and body structure of world-top judoists. Paper presented at the XI International Congress of Anthropological and Ethnological Sciences, Burnaby, Canada.

Farmosi, I. (1980). Body composition, somatotype and some motor performance of judoists. *Journal of Sports Medicine and Physical Fitness*, **20**, 431-434.

Hirata, K. (1966). Physique and age of Tokyo Olympic champions. *Journal of Sports Medicine and Physical Fitness*, **6**, 207-222.

Hirata, K. (1979). *Selection of Olympic champions*. Basel: Karger.

Maas, G.D. (1974). *The physique of athletes*. Leiden: Leiden University Press.

Matsumoto, Y., Ogawa, S., Asami, T., Ishiko, T., Kawamura, T., Daigo, T., Katsuta, S., Masuda, M., & Shibayama, H. (1969). Physical fitness of the top judoists in Japan—1967. In Association for the Scientific Studies on Judo (Ed.), *Bulletin of the Association for the Scientific Studies on Judo, Kodokan. Report III* (pp. 1-12). Tokyo: Kodokan.

Mery, J., & Marbach, G. (1979). Surveillance médicale de la section judo. Études du lycée technique nationalisé industrialisé de Strasbourg, Année 1976-1977. *Médecine du Sport*, **53**, 79-84.

Ostyn, M., Simons, J., Beunen, G., Renson, R., & van Gerven, D. (Eds.). (1980). *Somatic and motor development of Belgian secondary schoolboys*. Leuven: Leuven University Press.

Renson, R., Beunen, G., van Gerven, D., Simons, J., & Ostyn, M. (1980). Description of motor ability tests and anthropometric measurements. In M. Ostyn, J. Simons, G. Beunen, R. Renson, & D. van Gerven (Eds.), *Somatic and motor development of Belgian secondary schoolboys. Norms and standards* (pp. 24-44). Leuven: Leuven University Press.

Ross, W.D., Ward, R., Leahy, R.M., & Day, J.A.P. (1982). Proportionality of Montreal athletes. In J.E.L. Carter (Ed.), *Physical structure of Olympic athletes: Part I. The Montreal Olympic Games anthropological project: Vol. 16. Medicine and sport* (pp. 81-106). Basel: Karger.

Sheldon, W.H., Dupertuis, C.W., & McDermott, E. (1954). *Atlas of men. A guide for somatotyping the adult male at all ages*. Darien, CT: Hafner.

Stephens, W.G.S., & Taylor, J.H. (1962). The schematic two-dimensional plotting of the spatial relationships among somatotypes. *American Journal of Physical Anthropology,* **20**, 395-398.

Taylor, A.W., & Brassard, L. (1981). A physiological profile of the Canadian judo team. *Journal of Sports Medicine and Physical Fitness,* **21**, 160-164.

Toteva, M., & Slanchev, P. (1983, September). *Somatotype of top-class Bulgarian judo players.* Paper presented at the meeting of the International Congress on Sports and Health, Maastricht, The Netherlands.

15

The Anatomical and Physiological Characteristics of Preadolescent Swimmers, Tennis Players, and Noncompetitors

John Bloomfield, Brian A. Blanksby,
Timothy R. Ackland, and Bruce C. Elliott
UNIVERSITY OF WESTERN AUSTRALIA
NEDLANDS, WESTERN AUSTRALIA, AUSTRALIA

The longitudinal growth and development study presently being conducted at the University of Western Australia involves the measurement of approximately 120 elite age-group female and male swimmers, 120 elite junior female and male tennis players, and 220 noncompetitive female and male primary school children. The latter group was selected because they were not involved in regular formal training for sport more than once per week, and they matched the socioeconomic status of the swimmers and tennis players. All subjects are measured at 6-month intervals until they reach late adolescence.

Method

For the purposes of this study 202 preadolescent children rated by pubescent assessment as at stage 1 (Tanner, 1962) and between 7 and 12 years old were selected from the total sample. Of this selected group, 37 children were in the competitive swimming subpopulation, 61 in the competitive tennis subpopulation, and 104 in the noncompetitive group.

Administration of the Tests

Cable tension tests to measure the strength of thigh flexion, leg extension, and arm extension were carried out following the procedures developed by Clarke (1976), and handgrip strength was determined using a TEC Handgrip Dynamometer. Flexibility at the ankle, hip, and shoulder joints was assessed using a Leighton Flexometer (Leighton, 1942). Harpenden calipers were used to assess triceps, subscapular, suprailiac, and calf skinfolds. The sum of the triceps, subscapular, and suprailiac measures was subsequently used to provide an indication of body fat. The somatotype of each subject was assessed according to the Heath-Carter Method (Carter, 1980), and leg power was calculated using the results from a modified vertical jump and the Lewis Nomogram (Mathews & Fox, 1971). A Vitalograph Single Breath Wedge Spirometer was used to assess forced vital capacity (FVC) and forced expiratory volume in the first second (FEV_1), while a PWC_{170} test was employed to determine the physical exercise capacity of each subject on a Monark bicycle ergometer.

Analysis

A series of analyses of variance were applied to the data to determine whether there were significant differences between the three groups over each age and gender classification. Interactions between age, gender, and treatment were not significant, which indicated that treatment effects did not depend on gender and age. These interaction terms were therefore dropped from the model (Aitken, 1978). Similarly, when testing the significance of the main effects, if age and gender effects were significant for the variable in question, they were retained in the model. A significance level of .025 was adopted to compensate for the overall possibility of Type I error due to the multiple F-tests used in the analysis (Aitken, 1978).

For those variables where the F-ratio indicated a significant difference, t-tests were employed to determine where significant differences existed. To correct for the increased probability of obtaining a significant result due to multiple comparisons, the significance level (.05) was divided by the number of comparisons to be made, three in this case. This followed the recommendation of Miller (1956) for multiple comparisons.

Results

No significant differences were found between competitors and noncompetitors in height, body mass, or somatotype, which supports the findings of several previous studies (Andrew, Becklake, Guleria, & Bates, 1972; Cunningham & Enyon, 1973; Magel & Andersen, 1969; Sobolova, Seliger, Grissova, Machovcoca, & Zelenka, 1971). Body fat, as determined by the sum of three skinfold measures (triceps, subscapular, suprailiac), was also similar for all three groups. This result was in contrast to that of earlier studies (Sobolova et al., 1971; Vaccaro, Clarke, & Morris, 1980) which showed young swimmers to have lower than normal levels of fat deposition.

With respect to joint mobility, again this study revealed no significant differences in arm flexion-extension, thigh latero-medial rotation or foot dorsiplantar flexion between the three subpopulations. This was in contrast to studies of adult swimmers (Bloomfield, 1967; Cureton, 1940) and may be due to the high degree of flexibility at the shoulder, hip, and ankle joints of young children. Because joint flexibility in a normal population has been shown to decrease with age (Hupprich & Sigerseth, 1950), it would appear that given the results of Cureton (1940) and Bloomfield (1967), the dry land and in-the-water training undertaken by the swimming group may serve to maintain their level of shoulder joint mobility as they mature.

The results of the strength tests revealed that the three groups were similar in thigh flexion strength, while the swimmers exhibited significantly greater leg extension strength and arm extension strength than subjects in both the noncompetitive and tennis groups. The development of leg extension strength would improve the swimmer's performance not only while kicking but also during the starts and turns. One might have presumed that this capacity would also develop in tennis players due to the weight-supportive nature of the game and the requirement for rapid acceleration and changes of direction. Arm extension strength at the shoulder joint provides the majority of the propulsive force in all swimming strokes (Cureton, 1970; Piette & Clarys, 1979); however, it is also required in the serve and other tennis strokes. Therefore, one might have expected the tennis players to exhibit a higher score than the noncompetitors on this measure. Apparently, the repetition of this movement by the tennis players at this stage of development and level of training is insufficient to produce the same effect as the repetitive movement used in every swimming stroke.

No difference among the three groups was found on the handgrip strength score, which conflicts somewhat with the findings of Vodak, Savin, Haskell, and Wood (1980) and Powers and Walker (1982) for adolescent tennis players. The fact that the tennis players were not superior on this measurement could reflect the level of training experienced at this stage. The importance of handgrip strength in tennis has recently been reported by Elliott (1982), who showed that while the firmness with which one grips a tennis racquet has little influence on impacts at the center of percussion of the implement, a firm grip does provide an advantage in terms of rebound velocity for off center impacts. It was further found that a comparison of vertical jump and leg power scores revealed that no differences existed among the three groups.

The swimming group demonstrated superior FVC compared to the noncompetitors and tennis players. This was in agreement with findings from many previous studies (Åstrand et al., 1963; Magel & Andersen, 1969; Newman, Smalley, & Thomason, 1961; Vaccaro et al., 1980). This result is substantiated by Andrew et al. (1972) who controlled for the effects of height and found swimmers to have a superior FVC when compared to noncompetitors.

No significant difference was found between any groups in the measure of $FEV_1\%$. Given that the swimmers possessed a higher FVC, yet forcefully expired the same percentage of FVC in one second as the other two groups, a greater expiratory flow rate for the swimmers was inferred. This finding is supported by Andrew et al. (1972), who measured slightly older children who were not necessarily preadolescent.

Further examination of the data showed that the swimmers had superior physical exercise capacity scores compared to the tennis players and noncompetitors. This finding was in agreement with a number of studies that had reported higher oxygen uptake and PWC_{170} values in young swimmers compared to untrained children of the same ages (Åstrand et al., 1963; Cunningham & Enyon, 1973; Robinson et al., 1978; Sobolova et al., 1971). However, the similarity in scores between the tennis players and noncompetitors was not in agreement with the findings of Powers and Walker (1982), who found that adolescent tennis players possessed higher oxygen uptakes than noncompetitors.

A survey of the children prior to the test period revealed that the swimmers were averaging five training sessions per week. This training was formally structured, involving the development of cardiovascular fitness as well as skill. The preadolescent tennis players, however, were training only two to three times a week, and this was almost entirely skill oriented. With postadolescent tennis players, a comparison with noncompetitors would have been expected to indicate a difference in scores because of the increased training intensity. In view of the level of training by the preadolescent tennis players, however, the results obtained were not surprising.

Discussion

It would appear that at this preadolescent stage of development, physical size, body shape, body composition, and flexibility are not discriminating factors for high or low levels of performance, because the subjects in this study showed few definitive differences in those tests considered to be valuable in differentiating competitors from noncompetitors. It has been reported, however, that there are considerable size, body shape, strength, and flexibility differences in postadolescent males and females competing at a high level in swimming and tennis when they are compared to noncompetitors (Bloomfield & Blanksby, 1971; Elliott, Ducat, Ellis, & Anglim, 1982).

One question that arises is whether preadolescent children need the rigorous flexibility and strength training which some coaches feel is important for them at this stage. Few studies have examined the flexibility of children from the preadolescent through the postadolescent years. The limited information which does exist suggests that flexibility increases just prior to age 12 and decreases thereafter (Herkowitz, 1978; Hupprich & Sigerseth, 1950). The results of a longitudinal study of young swimmers by Butgakowa and Woroncow (1980) showed that the greatest development in flexibility was made during prepuberty and early puberty, prior to age 14. From approximately this age onwards, it would appear that dryland training is essential to maintain optimum levels of flexibility for swimmers.

The question of strength training programs for the preadolescent athlete has still to be satisfactorily resolved. Few studies have addressed the question of strength trainability in the preadolescent child, but the results of this study show swimmers to possess greater leg and arm extension strength than the noncompetitors, presumably as a result of regular training. Jones (1949) stated, however, that strength training for the prepubertal boy or girl was not very productive, as the gains were small until the androgenic hormones were produced in sufficient amounts to permit muscle hypertrophy. Therefore the value of strength training during preadolescence remains a subjective issue, and it may be more beneficial to spend extra time in the development of proper stroke mechanics at this stage (Shaffer, 1980).

A further point which arises from the findings is related to the effect of training on the swimmers. This group revealed higher FVC scores than either the tennis players or the noncompetitors, and at the same time had higher physical exercise capacities, which are an indication of endurance ability. There is some contention among swim coaches as to whether endurance training or intensive training over short distances with an emphasis on technique development is more beneficial for preadolescent children. While the principles of specificity of training are always important, it appears that preadolescent children can adapt to aerobic work in a similar fashion to adults (Eriksson, 1972). The maximal a-$\bar{v}O_2$ differences are also similar for children and adults (Eriksson, 1972, 1978), and some evidence suggests that training during preadolescence and adolescence produces a greater increase in the size of the cardiorespiratory system organs than does training in later life (Åstrand et al., 1963; Eriksson, 1972).

Inasmuch as several authorities have subjectively warned of the dangers inherent in the vigorous training of young children, the results of this study suggest, at least in terms of growth, that there were few differences between the groups, and normal growth patterns did not seem to be endangered. As the major differences were superior strength and cardiovascular endurance for the swimming group, their more intensive training program can only be considered beneficial.

References

Aitken, M. (1978). The analysis of unbalanced cross-classifications. *Journal of Royal Statistical Society,* **141**(2, Series A), 195-223.

Andrew, G.M., Becklake, M.R., Guleria, J.S., & Bates, D.V. (1972). Heart and lung functions in swimmers and non-athletes during growth. *Journal of Applied Physiology,* **32**(2), 245-251.

Åstrand, P.O., Engstrom, L., Eriksson, B.O., Karlberg, P., Nylander, I., Saltin, B., & Thoren, C. (1963). Girl swimmers. *Acta Pediatrica Scandinavica,* **42** (Supp. 147), 1-75.

Bloomfield, J. (1967). *Anatomical and physiological differences between three groups of swimmers of varying ability.* Unpublished doctoral dissertation, University of Oregon, Eugene.

Bloomfield, J., & Blanksby, B.A. (1971). Strength, flexibility and anthropometric measurements. A comparison of highly successful male university swimmers and normal university students. *The Australian Journal of Sports Medicine,* **3**(10), 8-15.

Butgakowa, N.Z., & Woroncow, A.R. (1980). Selection and training of young swimmers. *Sport Sypoczywowy* [Poland], **18**(1/2), 39-45.

Carter, J.E.L. (1980). *The Heath-Carter somatotype method.* San Diego: San Diego State University.

Clarke, H.H. (1976). *Application of measurement to health and physical education.* New Jersey: Prentice-Hall.

Cunningham, D.A., & Enyon, R.B. (1973). The working capacity of young competitive swimmers 10-16 years of age. *Medicine and Science in Sports*, **8**(4), 227-231.

Cureton, T.K. (1940). Review of a decade of research in aquatics at Springfield College. *Research Quarterly*, **11**(2), 68-79.

Cureton, T.K. (1970). Biomechanics of swimming with interrelationship to fitness and performance. In L. Lewillie & J.P. Clarys (Eds.), *Biomechanics in swimming* (pp. 31-52). Brussels: Universite Libre de Bruxelles.

Elliott, B.C. (1982). Tennis: The influence of grip tightness on reaction impulse and rebound velocity. *Medicine and Science in Sports and Exercise*, **14**(5), 348-352.

Elliott, B.C., Ducat, J., Ellis, S., & Anglim, J. (1982). Functional fitness for tennis. *Sports Coach*, **6**(1), 46-47.

Eriksson, B.O. (1972). Physical training, oxygen supply and muscle metabolism in 11-13 year old boys. *Acta Physiologica Scandinavica*, **87** (Supp. 384), 1-48.

Eriksson, B.O. (1978). Training girls for swimming from medical and physiological points of view, with special reference to growth. In B. Eriksson & B. Funberg (Eds.), *Swimming Medicine IV* (pp. 3-15). Baltimore: University Park Press.

Herkowitz, J. (1978). Sex-role expectations and motor behavior of the young child. In M.V. Ridenour (Ed.), *Motor development: Issues and applications* (pp. 32-40). New Jersey: Princeton Book Co.

Hupprich, F.L., & Sigerseth, P.O. (1950). Specificity of flexibility in girls. *Research Quarterly*, **21**(1), 25-33.

Jones, H.E. (1949). *Motor performance and growth: A developmental study of dynamometric strength*. Berkeley, CA: University of California Press.

Leighton, J.R. (1942). A single objective and reliable measure of flexibility. *Research Quarterly*, **13**(2), 206-212.

Magel, J.R., & Andersen, K.L. (1969). Pulmonary diffusing capacity and cardiac output in young well-trained Norwegian swimmers and untrained subjects. *Medicine and Science in Sports*, **1**(3), 131-139.

Mathews, D., & Fox, E. (1971). *The physiological basis of physical education and athletics*. Philadelphia: W.B. Saunders.

Miller, R.G. (1956). *Simultaneous statistical inference*. New York: McGraw-Hill.

Newman, E., Smalley, B.F., & Thomason, M.L. (1961). A comparison between body size and lung function of swimmers and normal school children. *Journal of Physiology*, **156**, 9-10.

Piette, G., & Clarys, J.P. (1979). Telemetric electromyography of the front crawl movement. In J. Terauds & E. Bedingfield (Eds.), *Swimming III. International Series on Sports Sciences* (Vol. 8). Baltimore: University Park Press.

Powers, S.K., & Walker, R. (1982). Physiological and anatomical characteristics of outstanding female junior tennis players. *Research Quarterly*, **53**(2), 172-175.

Robinson, P.S., Caffrey, G.P., Ridinger, R.R., Steinmetz, C.W., Bartels, R.L., & Shaffer, T.E. (1978). Metabolic effects of heavy physical training on female "age-group" swimmers. *British Journal of Sports Medicine*, **12**(1), 14-21.

Shaffer, T.E. (1980). The young athlete: New guidelines in sports medicine. *Pediatric Consultant*, **1**(5), 1-12.

Sobolova, V., Seliger, V., Grissova, D., Machovcoca, J., & Zelenka, V. (1971). The influence of age and sports training in swimming on physical fitness. *Acta Pediatrica Scandinavica*, (Supp. 217), 63-67.

Tanner, J.M. (1962). *Growth at adolescence* (2nd ed.). Oxford: Blackwell Scientific Publications.

Vaccaro, P., Clarke, D.H., & Morris, A.F. (1980). Physiological characteristics of young well-trained swimmers. *European Journal of Applied Physiology*, **44**, 61-66.

Vodak, P.A., Savin, W.M., Haskell, W.L., & Wood, P.D. (1980). Physiological profile of middle-aged male and female tennis players. *Medicine and Science in Sports and Exercise*, **12**(3), 159-163.

16

Body Size, Skinfolds, and Somatotypes of High School and Olympic Wrestlers

J.E. Lindsay Carter
SAN DIEGO STATE UNIVERSITY
SAN DIEGO, CALIFORNIA, USA

Frank D. Lucio
RAMONA HIGH SCHOOL
RAMONA, CALIFORNIA, USA

Studies on Olympic wrestlers support the concept of a relationship between physique and success in international competition (Carter, 1982; De Garay, Levine, & Carter, 1974; Kohlrausch, 1930; Tanner, 1964). In general, the Olympic wrestler is very mesomorphic with low endomorphy and ectomorphy. His average somatotype is 2.5-6.5-1.5, but there is some variation according to weight class. There have only been a few studies in North America on the physiques of high school wrestlers (Clarke, 1974; Katch & Michael, 1971; Tcheng & Tipton, 1973; Tipton & Tcheng, 1970), and only two studies have reported somatotypes (Sinning, Wilensky, & Meyers, 1976; Thorland, Johnson, Fagot, Tharp, & Hammer, 1981). None of these studies made comparisons between high school (HS) and Olympic (OL) wrestlers. An investigation of the differences or similarities between OL and HS wrestlers can improve understanding of the morphological characteristics that may be related to success of wrestlers at both levels.

This study compared high school and Olympic wrestlers on age, height, weight, sum of four skinfolds, and somatotype. It was hypothesized that there would be differences in size, skinfolds, and somatotype among weight group(s), and differences in size, skinfolds, and somatotype between HS and OL wrestlers by weight group.

Studies on Olympic wrestlers support the concept of a relationship between physique and success in international competition (Carter, 1982; De Garay, Levine, & Carter, 1974; Kohlrausch, 1930; Tanner, 1964). In general, the Olympic wrestler is very mesomorphic with low endomorphy and ectomorphy. His average somatotype is 2.5:6.5:1.5, but there is some variation according to weight class. There have only been a few studies in North America on the physiques of high school wrestlers (Clarke, 1974; Katch & Michael, 1971; Tcheng & Tipton, 1973; Tipton & Tcheng, 1970), and only two studies have reported somatotypes (Sinning, Wilensky, & Meyers, 1976; Thorland, Johnson, Fagot, Tharp, & Hammer, 1981). None of these studies made comparisons between high school (HS) and Olympic (OL) wrestlers. An investigation of the differences or similarities between OL and HS wrestlers can improve understanding of the morphological characteristics that may be related to success of wrestlers at both levels.

This study compared high school and Olympic wrestlers on age, height, weight, sum of four skinfolds, and somatotype. It was hypothesized that there would be differences in size, skinfolds, and somatotype within weight group, and differences in size, skinfolds, and somatotype between HS and OL wrestlers by weight group.

Methods

The HS subjects ($N = 83$) were wrestlers who qualified for and competed in the 1978 California State High School Wrestling Championships held at San Diego State University, San Diego, California. The OL wrestlers ($N = 78$) were freestyle and Greco-Roman wrestlers who competed at the 1968 Olympic Games in Mexico City (De Garay et al., 1974). As there were no physique differences between the two styles, they were combined for this study. Because the range of weights was different for HS and OL wrestlers, it was necessary to establish equivalent weight classifications to accommodate both samples of wrestlers and to provide adequate numbers in each classification. Weight on the day of measurement was used for grouping, rather than weight class in which the wrestler usually competed.

The following anthropometric dimensions were taken following the procedures of De Garay et al. (1974) and Carter (1980): age, height, weight, four skinfolds (triceps, subscapular, anterior suprailiac, medial calf), two bone breadths (biepicondylar humerus and femur), and two girths (flexed and tensed upper arm, and maximal calf). The sum of four skinfolds (SUM4) was used as an indicator of subcutaneous fatness, and the somatotype was estimated by the Heath-Carter anthropometric method (Carter, 1980). The procedures used for somatotype analysis were as described in Carter (1980) and Carter, Ross, Duquet, and Aubry (1983). Means for independent variables were compared using t-tests or ANOVA, with $p < .05$.

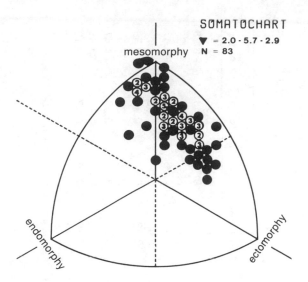

Figure 1. Somatoplots of California high school wrestlers.

Results

High School Wrestlers

Descriptive statistics for the HS wrestlers by weight group are presented in Table 1, and their somatotypes are plotted in Figure 1. The test-retest reliabilities for the anthropometric variables ranged between $r = .93$ and $r = .99$. Most of the subjects were 17 or 18 years old with a slight trend to increased age with weight group. Logically, height and weight increased with weight group. Between-group comparisons on SUM4 showed that the three lightest groups did not differ from each other, but they had smaller SUM4 than the two heaviest groups, and the latter two differed from each other. The scatter of somatotypes, or somatotype attitudinal mean (SAM), was similar for all groups, but the mean somatotypes for the two heavier groups (they were more endomesomorphic) differed from the three lighter groups.

Olympic Wrestlers

Mean age increased about 2.5 years from the lightest to heaviest weight group, but the standard deviations were approximately 4 to 6 years (see Table 2 and Figure 2). As expected, height and weight increased with weight group. There were no significant differences among the SAMs, but the heaviest groups differed in somatotype (more endomesomorphy) from the lightest group.

Table 1. Descriptive statistics for California high school wrestlers by five weight groups

Weight group (kg)		Age (yr)	Ht (cm)	Wt (kg)	HWR[a]	Somatotype			SAM[b]	SUM4[c]
52.2-58.9	M	17.56	168.0	55.1	44.15	1.7	4.9	3.7	0.94	23.8
(n = 18)	SD	1.00	3.53	2.19	1.03	0.35	0.66	0.79	0.48	3.53
59.0-63.9	M	17.69	170.70	59.8	43.70	1.8	5.4	3.4	1.05	24.2
(n = 16)	SD	0.49	4.70	4.88	1.29	0.31	0.74	0.96	0.62	4.47
64.0-71.1	M	17.76	174.46	66.3	43.11	1.8	5.4	3.0	1.06	24.7
(n = 15)	SD	0.81	4.43	2.07	1.29	0.36	0.93	0.94	0.81	3.58
71.2-80.6	M	17.93	176.41	72.4	42.33	2.1	6.2	2.4	1.05	27.5
(n = 20)	SD	0.69	4.97	2.55	1.01	0.43	0.83	0.79	0.56	4.47
80.7-88.0	M	18.01	181.70	81.0	42.00	2.7	6.5	2.2	1.14	34.8
(n = 14)	SD	0.47	4.71	4.63	0.89	0.75	0.86	0.70	0.60	9.10

[a]HWR = height/cube root of weight; [b]SAM = somatotype attitudinal mean; [c]SUM4 = sum of triceps, subscapular, anterior suprailiac, and medial calf skinfolds.

Table 2. Descriptive statistics for Olympic wrestlers at Mexico City by five weight groups

Weight group (kg)		Age (yr)	Ht (cm)	Wt (kg)	HWR[a]	Somatotype			SAM[b]	SUM4[c]
52.2-58.9	M	24.13	158.4	53.6	42.41	1.6	5.7	2.3	1.25	21.1
(n = 8)	SD	4.09	4.93	1.69	1.44	0.17	1.13	1.06	0.76	1.80
59.0-63.9	M	24.67	163.0	59.5	42.14	1.8	6.1	2.0	1.24	23.6
(n = 15)	SD	3.98	5.15	1.84	1.21	0.55	1.03	0.91	0.71	4.72
64.0-71.1	M	26.64	166.9	65.3	41.85	1.8	6.2	1.8	0.83	24.5
(n = 14)	SD	5.89	3.93	1.35	0.85	0.42	0.63	0.69	0.54	4.76
71.2-80.6	M	26.89	169.3	71.5	41.17	2.5	6.4	1.4	1.20	29.7
(n = 27)	SD	5.71	3.37	2.17	0.89	1.00	0.80	0.53	0.65	9.49
80.7-88.0	M	26.57	177.3	82.2	41.09	2.8	6.9	1.3	1.28	31.0
(n = 14)	SD	4.33	5.01	3.41	1.22	0.78	1.12	0.66	0.70	7.14

[a]HWR = height/cube root of weight; [b]SAM = somatotype attitudinal mean; [c]SUM4 = sum of triceps, subscapular, anterior suprailiac, and medial calf skinfolds.

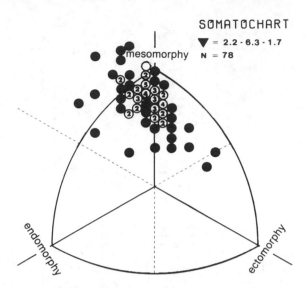

SOMATOCHART

▼ = 2.2 - 6.3 - 1.7

N = 78

mesomorphy

endomorphy

ectomorphy

Figure 2. Somatoplots of Olympic wrestlers (Mexico City, 1968).

High School Versus Olympic Wrestlers

The two samples did not differ in weight for any group, indicating that our sampling effectively controlled for weight differences. Within each weight group HS wrestlers were younger by 6 to 9 years, taller by 4.4 to 9.6 cm, and were higher on height-weight ratios (HWR) and ectomorphy (see Table 3 and Figure 3). HS wrestlers were less mesomorphic than OL wrestlers in the three lightest weight groups but did not differ in the two heaviest groups. Furthermore, the two samples did not differ on SUM4 or endomorphy in any weight group. When the total samples were compared in terms of somatotype, they were similar on SAMs, but the mean somatotypes differed. Comparisons of the separate components revealed that they did not differ on endomorphy, but OL wrestlers were more mesomorphic and less ectomorphic than HS wrestlers.

Discussion

The present HS sample was about 1 year older than other North American HS samples. All HS samples were comparable in height and weight and in somatotype to the Californians in this study when allowance was made for different weight ranges. In most of the somatoplots for each weight group in this study, there was more variation on mesomorphy and ectomorphy than on endomorphy. Overall, the somatotypes of the heavier HS wrestlers were closer

Table 3. Mean differences and t-ratios for high school versus Olympic wrestlers by weight group

Weight group (kg)		Age (yr)	Ht (cm)	Wt (kg)	HWR[a]	Somatotype $S_1 - S_2$	$SAM_1 - SAM_2$[b]	SUM4[c]
52.2-58.9	MD	-6.6	9.6	1.5	1.74	1.53	-0.3	2.8
(df = 24)	t	-6.54**	5.66**	1.72	3.52**	3.12**	1.19	2.09
59.0-63.9	MD	-7.0	7.0	0.0	1.56	1.57	-0.2	0.6
(df = 29)	t	-6.97**	4.34**	0.20	3.47**	3.17**	0.79	0.36
64.0-71.1	MD	-8.9	7.5	1.1	1.26	1.44	0.2	0.2
(df = 27)	t	-5.79**	4.81**	1.64	3.08**	3.10**	0.89	0.11
71.2-80.6	MD	-9.0	7.1	1.0	1.16	1.10	0.2	-2.2
(df = 45)	t	-6.9**	5.85**	1.42	4.17**	2.76**	0.81	-0.95
80.7-88.0	MD	-8.6	4.4	-1.1	0.91	0.99	0.1	3.9
(df = 26)	t	-7.35**	2.42*	-0.73	2.26*	1.79	0.56	1.25

[a]HWR = height/cube root of weight; [b]SAM = somatotype attitudinal mean; [c]SUM4 = sum of triceps, subscapular, anterior suprailiac, and medial calf skinfolds;
*$p < .05$; **$p < .01$.

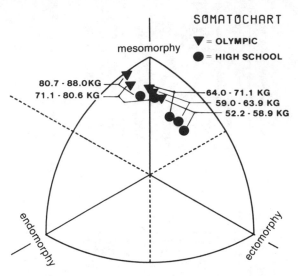

SOMATOCHART

▼ = OLYMPIC
● = HIGH SCHOOL

mesomorphy

80.7 - 88.0KG
71.1 - 80.6 KG

64.0 - 71.1 KG
59.0 - 63.9 KG
52.2 - 58.9 KG

endomorphy

ectomorphy

Figure 3. Mean somatoplots for Olympic and high school wrestlers by five weight groups.

to those of the heavier OL wrestlers than were the somatotypes of the lighter wrestlers.

The results show that there are obvious differences between HS and OL wrestlers in age, height, and somatotype for the total samples as well as for each weight group. Across weight groups HS wrestlers are 4.4 to 9.6 cm taller than their OL counterparts. This means that for the same weight OL wrestlers are much heavier than HS wrestlers per unit of height, and this is also reflected in lower HWRs and ectomorphy. The greater stockiness of OL wrestlers is accounted for by increased mesomorphy. Because more mesomorphy means more musculo-skeletal development relative to height, the OL wrestlers have greater relative muscle size and bone breadths. Although the maturity status of the HS wrestlers was not assessed, most are likely to be close to their maximal stature and some may be early maturers. However, anabolic processes should continue for some years and, in conjunction with continued training, could result in greater mass of muscle and bone tissue and thus greater mesomorphy. This indicates that for the HS wrestler to have a physique similar to that of the Olympians, he will have to increase mesomorphy (and therefore weight) to that of the equivalent Olympic weight group for present height of the HS wrestlers.

Within the limitations of this study, the following conclusions are warranted.

1. High school and Olympic wrestlers increased in height, sum of four skinfolds, and mesomorphy, and decreased in ectomorphy as weight group increased.

2. High school and Olympic wrestlers differed in age, height, mesomorphy, and ectomorphy, but were similar in skinfolds and endomorphy by weight groups.

3. These findings support the hypotheses of differences in size, skinfolds, and somatotypes with weight class and differences in age, size, and somatotypes between high school and Olympic wrestlers.

References

Carter, J.E.L. (1980). *The Heath-Carter somatotype method.* San Diego: San Diego State University Press.

Carter, J.E.L. (Ed.). (1982). *Physical structure of Olympic athletes: Part I. The Montreal Olympic Games anthropological project: Vol. 16. Medicine and sport.* Basel: Karger.

Carter, J.E.L., Ross, W.D., Duquet, W., & Aubry, S.P. (1983). Advances in somatotype methodology and analysis. *Yearbook of Physical Anthropology, 26,* 193-213.

Clarke, K.C. (1974). Predicting certified weight of young wrestlers: A field study of the Tcheng-Tipton method. *Medicine and Science in Sports, 6,* 52-57.

De Garay, A.L., Levine, L., & Carter, J.E.L. (1974). *Genetic and anthropological studies of Olympic athletes.* New York: Academic Press.

Katch, F.I., & Michael, E.D. (1971). Body composition of high school wrestlers according to age and wrestling weight category. *Medicine and Science in Sports, 3,* 190-194.

Kohlrausch, W. (1930). Zusammenhange von Korperform und Leistung. Ergebnisse der anthropometrischen Messungen an den Athleten der Amsterdamer Olympiade. *Artbeitsphysiologie, 2,* 187-204.

Sinning, W.E., Wilensky, N.F., & Meyers, E.J. (1976). Post-season body composition changes and weight estimation in high-school wrestlers. In J. Broekhoff (Ed.), *Physical education, sports and the sciences* (pp. 137-153). Eugene: Microform Publications.

Tanner, J.M. (1964). *The physique of the Olympic athlete.* London: Allen & Unwin.

Tcheng, T.K., & Tipton, C.M. (1973). Iowa wrestling study: Anthropometric measurements and the prediction of a "Minimal" body weight for high school wrestlers. *Medicine and Science in Sports, 5,* 1-10.

Thorland, W.G., Johnson, G.O., Fagot, T.G., Tharp, G.D., & Hammer, R.W. (1981). *Medicine and Science in Sports and Exercise, 13,* 332-338.

Tipton, C.M., & Tcheng, T.K. (1970). Iowa wrestling study: Weight loss in high school students. *Journal of the American Medical Association, 214,* 1269-1274.

17

The Puerto Rican Athlete Kinanthropometry Project: Age Group and Senior Wrestlers

Miguel A. Rivera, Miguel A. Albarrán,
Ruben D. Malavé, and Walter R. Frontera
SPORTS MEDICINE CLINIC
SANTURCE, PUERTO RICO

The evaluation of an individual's sports potential includes the analysis of various mechanical, physiological, psychological, and anthropometric factors. In technologically advanced nations, these factors have been studied extensively in both the athletic and general populations (American College of Sports Medicine, 1978; Carter, 1984; Correnti & Zauli, 1964; de Garay, Levine, & Carter, 1974; Kobayashi et al., 1978; Shephard, 1971, 1976; Tanner, 1964). In most industrialized countries the development of centers of higher education, technical personnel, and the application of sophisticated equipment to the study of human performance and/or sport science has produced valid and reliable information in the area of physique and performance. Unfortunately, this has not been the case for many of the developing countries. Recently, the Olympic Solidarity Committee of the International Olympic Committee has identified the Caribbean as an area of interest for the development of the sport sciences. In response to this necessity, various groups have united their efforts and expressed their interest in the sport sciences. As a contribution to the sport sciences movement in the Caribbean, the Puerto Rican Athlete Kinanthropometry Project (PRAKP) has been initiated. This is the first report of the PRAKP. This study determined and compared the somatotypes of elite Puerto Rican age group and National Senior Team wrestlers.

The authors would like to thank Dr. J.E.L. Carter for his assistance and contribution in the preparation of this article.

Method

Age group wrestlers ($N = 114$) and National Senior Team wrestlers ($N = 6$) were evaluated during the 1983 Puerto Rican Infantile and Juvenile National Championships and the 1983 Pan-American Games trials. The subjects (S) were grouped by age according to the International Amateur Wrestling Federation standards into the following categories (C): C1 6 to 8 years ($n = 4$), C2 9 to 11 years ($n = 25$), C3 12 to 13 years ($n = 26$), C4 14 to 15 years ($n = 36$), C5 16 to 18 years ($n = 24$), and C6 National Senior Team ($n = 6$). Age (years), height (cm), weight (kg), and the measures required for the Heath-Carter anthropometric somatotype (Carter, 1980) were taken. Within each category, means and standard deviations for each of the measures and for the somatotype components were calculated. An ANOVA was conducted on these means. The scatter of the somatotypes about their means was also examined by an ANOVA on the somatotype attitudinal mean (SAM) (Carter, 1980), as were the differences among the mean somatotypes as a whole.

Results

Table 1 shows the means and standard deviations of the variables by category. This table indicates that the Ss showed an increase in weight and height as a function of age ($p < .001$). An examination of differences between means for height and weight using the Tukey Multiple Comparison Test ($\alpha_{EW} = .05$) (Tables 2 & 3) showed significant differences in most cases. In the somatotype analysis (see Figure 1), a predominance of the mesomorphic component was observed across all categories. No significant differences ($p > .05$) for the SAM were noted.

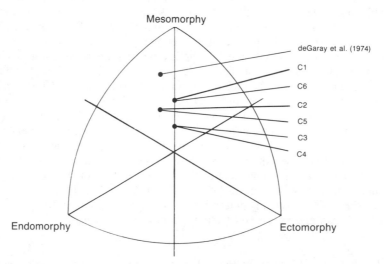

Figure 1. Somatochart for the 6 wrestling categories and the Olympic wrestlers of de Garay, Levine, and Carter (1974).

Table 1. Means and standard deviations of physical and somatotype characteristics of Puerto Rican wrestlers by age group

Variable	C1 (6 to 8) n = 4		C2 (9 to 11) n = 25		C3 (12 to 13) n = 26		C4 (14 to 15) n = 36		C5 (16 to 18) n = 24		C6 (19+) n = 6	
	M	SD	M	SD	M	SD	M	SD	M	SD	M	SD
Age	7.70	±0.97	10.64	±0.88	12.98	±0.56	15.03	±0.57	17.32	±0.83	20.14	±3.09
Weight (kg)	30.34	±7.60	34.88	±7.08	44.45	±11.18	57.95	±12.65	68.16	±13.10	71.24	±12.20
Height (cm)	130.97	±9.19	135.98	±7.18	149.17	±10.96	163.46	±6.02	167.66	±5.17	168.40	±8.19
Endomorphy	1.75	±0.75	3.08	±1.45	2.77	±1.77	2.60	±1.50	3.29	±1.62	2.33	±0.99
Mesomorphy	4.63	±0.65	5.15	±0.73	4.94	±1.04	4.74	±1.37	5.39	±1.81	5.33	±0.49
Ectomorphy	2.36	±0.41	2.26	±1.22	2.85	±1.51	2.67	±1.12	1.90	±1.30	1.91	±1.75
SAM[a]	1.06		1.81		2.12		1.94		2.51		1.37	

[a]SAM = somatotype attitudinal mean.

Table 2. Results of Tukey Multiple Comparison Test on mean weight (kg) scores and level of significance for Puerto Rican age group wrestlers

Age group and weight (kg)	C1 30.34	C2 34.88	C3 44.45	C4 57.95	C5 68.16	C6 71.24
C1 30.34	—	—	—	—	—	—
C2 34.88	NS	—	—	—	—	—
C3 44.45	NS	.05	—	—	—	—
C4 57.95	.05	.05	.05	—	—	—
C5 68.16	.05	.05	.05	.05	—	—
C6 71.24	.05	.05	.05	NS	NS	—

Note. C1 = 6 to 8 years; C2 = 9 to 11 years; C3 = 12 to 13 years; C4 = 14 to 15 years; C5 = 16 to 18 years; C6 = 19+ years (National Senior Team); NS = $p > .05$.

Table 3. Tukey Multiple Comparison Test mean height (cm) scores and level of significance for Puerto Rican age group wrestlers

Age group and weight (kg)	C1 130.97	C2 135.98	C3 149.17	C4 163.46	C5 167.66	C6 168.40
C1 130.97	—	—	—	—	—	—
C2 135.98	NS	—	—	—	—	—
C3 149.17	.05	.05	—	—	—	—
C4 163.46	.05	.05	.05	—	—	—
C5 167.66	.05	.05	.05	NS	—	—
C6 168.40	.05	.05	.05	NS	NS	—

Note. C1 = 6 to 8 years; C2 = 9 to 11 years; C3 = 12 to 13 years; C4 = 14 to 15 years; C5 = 16 to 18 years; C6 = 19+ years (National Senior Team); NS = $p > .05$.

Discussion

The results obtained for the height and weight variables were expected. Many other investigators have reported similar size differences among age groups (Wilton, 1955; Tanner, 1962). These differences are usually attributed to maturational factors. The consistency of the somatotypes across all categories can be attributed to the selection process. Presumably, the balanced mesomorphic somatotype shown by these wrestlers is a minimum qualification for enough success to allow participation in the national championship. The differences among the various somatotype attitudinal means were nonsignificant, again indicating a high degree of homogeneity of physique among the six groups of wrestlers studied. Considering that the subjects in this study included children and adolescents, these results are comparable to those of de Garay et al. (1974). In conclusion, these data suggest that a mesomorphic predominance is a requirement for successful participation in wrestling.

References

American College of Sports Medicine. (1978). Position statement on the quantity and quality of exercise for developing and maintaining fitness in healthy adults. *Medicine and Science in Sports, 3*, 7-10.

Carter, J.E.L. (1980). *The Heath-Carter somatotype method*. San Diego: San Diego State University Press.

Carter, J.E.L. (Ed.). (1984). *Physical structure of Olympic athletes: Part I. The Montreal Olympic Games anthropological project: Vol. 16. Medicine and sport*. Basel: Karger.

Carter, J.E.L. (Ed.). (1984). *Physical structure of Olympic athletes: Part II. Kinanthropometry of Olympic athletes: Vol. 16. Medicine and sport*. Basel: Karger.

Correnti, V., & Zauli, P. (1964). *Olimpionici 1960*. Rome: Marves.

de Garay, A.L., Levine, L., & Carter, J.E.L. (Eds.). (1974). *Genetic and anthropological studies of Olympic athletes*. New York: Academic Press.

Kobayashi, K., Kitamura, K.M., Miura, M., Sudeyama, H., Murase, Y., Miyashita, M., & Matsue, H. (1978). Aerobic power as related to body growth and training in Japanese boys: A longitudinal study. *Journal of Applied Physiology, 44*, 666-672.

Shephard, R.J. (Ed.). (1971). *Frontiers of fitness*. Springfield: Thomas.

Shephard, R.J. (1976). The work capacity of selected populations. In J. Weiner (Ed.), *International biological programme* (Vol. 4). Cambridge: University Press.

Tanner, J.M. (1962). *Growth at adolescence* (2nd ed.). Oxford: Blackwell.

Tanner, J.M. (1964). *The physique of the Olympic athlete*. London: George Allen and Unwin.

Wilton, M.K. (1955). The physical growth of children: An appraisal of studies 1950-1955. *Monographs of the Society for Research in Child Development, 20*(No. 1).

18

The Relationship of Somatotype and Body Composition to Strength in a Group of Men and Women Sport Science Students

Peter Bale
BRIGHTON POLYTECHNIC
EAST SUSSEX, ENGLAND

In recent studies by the author (Bale, 1979, 1980), a significant relationship was discovered between the physiques of women physical education students and their performance in motor activities involving speed, agility, and strength. The performance variables which correlated most highly with mesomorphy were grip strength, arm strength, an overall strength index, and 60 m dash. Moderate correlations were also obtained between lean body weight and grip strength and between lean body weight and mesomorphy. This study examined the relationship of somatotype and body composition to strength in a group of men and women sport science students aged 19 to 22 years.

One hundred and eleven students, 74 men and 37 women, in their second year of a bachelor of science (honors) sport science course were measured anthropometrically at a total of 10 sites. Biceps, triceps, subscapular, suprailiac, thigh, and calf skinfolds were measured using Harpenden calipers with a caliper tension of 20.3 gmm^{-2}. Total skinfold was calculated from the sum of five of these six skinfold measurements. The anterior thigh skinfold was omitted because of the difficulty in accurately measuring this skinfold on these athletic students. Bicondylar measurements were made using the Harpenden Anthropometer and circumferences measured with a constant tension steel tape. Height and weight were also measured. All the skinfolds were taken in a standing position except for the biceps fold, which was taken with the arm resting on a table and the hand supinated. The Heath-Carter method of estimating somatotype was selected (Heath & Carter, 1967).

Initially, body density was estimated from the logarithm of the sum of four skinfolds using the regression equations suggested by Durnin and Rahaman (1967) and Durnin and Womersley (1974). However, because regression equations for predicting body composition from anthropometric measurements are considered to be population-specific (Fleck, 1983; Katch & Katch, 1980; Jackson, Pollock, & Ward, 1980; Sinning, 1980; Wilmore & Behnke, 1970), the accuracy of the regression equations used was checked by hydrostatic weighing (Harsha, Fredrichs, & Berenson, 1978). Residual volume was calculated from analysis of percent oxygen and carbon dioxide (Wilmore, Vodak, Parr, Girandola, & Billing, 1980) using a Beckman OM-11 O_2 analyzer and a Morgan infrared CO_2 analyzer.

The densities of 60 of the students were measured using hydrostatic weighing and these densities correlated with those obtained by the regression analysis. The correlation of .96 confirmed the accuracy of the regression analysis used for the men students, but the regression analysis used to estimate the densities of the women students overestimated them by 1 to 2.5%; the correlation coefficient was only .83. When a paired t-test was used, there was no difference between the densities estimated by multiple regression analysis and by underwater weighing in the men students, but the t value of -4.92 for the women students was significant at the .01 level of confidence. New regression equations were therefore created using the skinfold measurements and the densities obtained by hydrostatic weighing of these 60 students. On the basis of these comparisons the following new multiple regression equations were used to calculate the densities of all the students in the study.

Males: Density = 1.1559 − 0.0583X
Females: Density = 1.1297 − 0.0488X
X = log of the sum of the biceps, triceps, subscapular, and suprailiac skinfolds.

Percent fat, absolute fat, and lean body weight were estimated from body density using the Siri equation (1956). This equation assumes that lean body mass has a constant density of 1.10 gm/ml^{-1}, and it is considered by its author to have a high reliability and to accurately estimate body composition compared to direct methods.

A Harpenden hand dynamometer was used to test the grip strength of both hands and a Takei Kiki Kogyo dynamometer was used to measure back strength and leg lift. Arm strength of both arms was measured using an arm pull apparatus connected to a force transducer; this apparatus was developed in our laboratory. The index of total strength was calculated by adding the six strength scores. Product-moment correlations between the physique, body composition, and strength variables were calculated. Discriminant analysis was also carried out to search for significant differences in physique and performance between the men and women students. Finally, these students were divided into sport groups according to the specialty pursued by each student while in the sport science course. As a result, nine sport groups were identified for the men and five groups for the women.

The somatotypes of the male sport science students are all found in the endomorphic mesomorph, ectomorphic mesomorph, and mesomorphic ectomorph sectors of the somatochart, with 80% of the somatotypes confined to the predominantly muscular builds. The somatotypes of the women students are found mainly in the endomorphic mesomorph and mesomorphic endomorph sectors of the somatochart, but they are less mesomorphic and more endomorphic than their male counterparts. However, both the men and women students who are active in sport are generally taller, heavier, more muscular, and less fat than nonathletic students of the same age (Carter, 1970, 1981; de Garay, Levine, & Carter, 1974; Parnell, 1958). The somatotypes of most women students, for example, lie below the ectomorphic axis in the endomorphic sectors of the somatochart. The somatotypes of nonathletic men students are found all over the chart with most somatotypes centered around the balanced mesomorph areas.

The mean values and standard deviations of the physique and body composition measurements of the students are shown in Table 1. The body composition measures are very close to the mean values reported by Katch and Michael (1968) and by Sloan, Burt, and Blyth (1962) for women students, and by Sloan (1967) for male students. However, comparison of the present data with body composition values reported by Wilmore and Behnke (1968, 1970), Damon and Goldman (1964), and by Slaughter and Lohman (1976) supports the view that these sport science students have higher densities, less fat, and higher lean body weights for their age than unselected students. Indeed, the sport science students in this study were similar in body composition to many of the athletes in various sports studied and reported by Desipre, Van Reenen, Gelderblom, and Barnard (1976), Fleck (1983), Forsyth and Sinning (1973), and Wilmore (1983).

The means and standard deviations of the scores on the strength measures are presented in Table 2. The mean grip strength scores by the men students in the present study are similar to those achieved by students tested by Montoye and Lamphiear (1977). The grip strengths of the women students are superior to Montoye's and Lamphiear's women students of a similar age and also to outstanding tennis players measured by Powers and Walker (1982). However, they are lower than the grip strength scores achieved by the women athletes studied by Heyward and McCreary (1977) and by physical education students measured by Bale (1980). As strength is related to cross-sectional area of muscle to some degree, it is expected that a larger, more muscular athlete would have greater strength. Thus, when differences in physique, body composition, and strength between the men and women students in this study were examined, the men students were significantly taller, heavier, more muscular, less fat, and stronger than the women students.

In order to examine the relationship among physique, body composition, and strength variables, zero-order correlations were computed (Tables 3 and 4). In an earlier study of women physical education students (Bale, 1980), the correlation between mesomorphy and lean body weight was .43. However, in this study, even lower relationships between mesomorphy and lean body weight were obtained ($r = .40$). Slaughter and Lohman (1976) and Slaughter, Lohman, and Boileau (1977) also found low relationships between these two components in their studies of women students. They comment that this finding is not surprising because mesomorphy is defined as lean body weight in relation to height rather than as absolute lean body weight. Much higher rela-

Table 1. Means and standard deviations of physique and body composition measurements of men and women sport science students

Subjects		Age (yrs)	Height (cm)	Weight (kg)	Endo-morphy	Meso-morphy	Ecto-morphy	Ponderal index	Total skinfold (mm)	Density	% fat	Absolute fat	Lean body weight (kg)
Men	M	20.4	179.8	73.3	2.3	4.7	3.3	13.1	31.2	1.0752	10.4	7.8	65.6
	SD	1.49	5.9	7.4	0.7	0.9	1.1	0.4	9.0	0.0065	2.8	2.7	5.6
	n	74	74	74	74	74	74	74	74	74	74	74	74
Women	M	20.0	165.2	60.5	3.6	4.2	2.7	12.8	52.3	1.0522	20.4	12.5	48.0
	SD	1.27	5.2	7.4	0.8	0.8	1.1	0.5	14.4	0.0061	2.7	2.8	5.0
	n	37	37	37	37	37	37	37	37	37	37	37	37
F value			163.0**	74.1**	69.5**	8.9**	6.7*	8.4**	88.9**	323.0**	326.6**	74.4**	260.5**

$*p < .05; **p < .01.$

Table 2. Means and standard deviations of strength measurements of men and women sport science students

Subjects		Right grip (kg)	Left grip (kg)	Right arm strength (kg)	Left arm strength (kg)	Back strength (kg)	Leg lift (kg)	Overall strength (kg)
Men	M	51.7	47.3	25.3	23.9	136.1	184.9	468.3
	SD	7.4	7.1	4.9	4.9	24.0	31.5	61.9
	n	73	73	73	72	66	69	67
Women	M	36.5	32.4	16.8	15.5	96.2	122.6	321.7
	SD	6.3	5.6	2.5	2.8	16.6	20.6	40.7
	n	37	37	37	37	36	34	34
F value		114.1**	124.2**	98.5**	92.7**	78.8**	109.2**	155.4**

$**p < .01.$

Table 3. Zero-order coefficients for physique, body composition, and strength of the men sport science students[a]

Variable	1	2	3	4	5	6	7	8	9	10	11	12	13	14	15	16	17	18
1. Height		.54	-.09	-.27	.25	.30	-.05	.03	-.03	.13	.65	.26	.26	.14	.19	.29	.17	.29
2. Weight			.33	.41	-.55	-.55	.53	-.61	.61	.77	.95	.48	.45	.49	.43	.44	.48	.61
3. Endomorphy				.07	-.19	-.32	.81	-.79	.79	.73	.09	-.12	-.08	-.04	-.08	.01	.19	.10
4. Mesomorphy					-.74	-.71	.25	-.29	.29	.35	.37	.39	.27	.34	.18	.37	.47	.50
5. Ectomorphy						.96	-.46	.54	-.54	-.58	-.45	-.35	-.29	-.44	-.32	-.23	-.36	-.42
6. Ponderal index							-.51	.59	-.59	-.63	-.42	-.29	-.26	-.40	-.28	-.26	-.34	-.41
7. Total skinfold								-.95	.95	.91	.10	.10	.13	.10	.03	.01	.29	.21
8. Density									-1.00	-.97	-.34	-.14	-.15	-.17	-.10	-.04	-.32	-.26
9. Percent fat										.97	.34	.14	.15	.17	.10	.04	.32	.26
10. Absolute fat											.54	.25	.24	.25	.19	.15	.39	.38
11. Lean body weight												.52	.48	.53	.49	.51	.44	.64
12. Right grip strength													.83	.51	.48	.49	.54	.76
13. Left grip strength														.50	.56	.53	.36	.70
14. Right arm strength															.76	.41	.35	.59
15. Left arm strength																.39	.34	.57
16. Back strength																	.42	.79
17. Leg lift																		.83
18. Overall strength																		

Note. When $r > .22$, $p < .05$; when $r > .29$, $p < .01$; [a]$N = 74$.

Table 4. Zero-order coefficients for physique, body composition, and strength of the women sport science students[a]

Variable	1	2	3	4	5	6	7	8	9	10	11	12	13	14	15	16	17	18
1. Height		.56	-.04	-.23	.18	.23	.03	-.06	.06	.33	.64	.22	.13	.29	.15	.02	.10	.11
2. Weight			.59	.40	-.64	-.65	.63	-.67	.67	.91	.97	.42	.30	.17	.13	.13	.25	.38
3. Endomorphy				.23	-.63	-.70	.87	-.89	.89	.82	.42	.01	-.03	-.01	-.11	-.22	-.08	.05
4. Mesomorphy					-.65	-.70	.26	-.25	.24	.34	.40	.31	.30	-.05	.13	.40	.33	.47
5. Ectomorphy						.93	-.74	.74	-.73	-.74	-.53	-.37	-.29	.16	.05	-.05	-.21	.28
6. Ponderal index							-.76	.71	-.71	-.75	-.54	-.27	-.25	.15	.07	-.09	-.27	.39
7. Total skinfold								-.94	.95	.88	.44	-.05	-.11	-.12	-.20	-.24	-.03	-.07
8. Density									-1.00	-.91	-.47	-.10	-.02	-.09	-.13	-.32	-.07	-.09
9. Percent fat										.91	.47	.10	.01	.02	-.02	-.11	.11	.17
10. Absolute fat											.78	.25	.14	.24	.20	.26	.31	.47
11. Lean body weight												.48	.37	.27	.33	.36	.37	.63
12. Right grip strength													.81	.24	.34	.29	.40	.60
13. Left grip strength															.42	.50	.10	.41
14. Right arm strength																.41	.08	.39
15. Left arm strength																	.53	.83
16. Back strength																		.85
17. Leg lift																		
18. Overall strength																		

Note. When $r > .32$, $p < .05$; when $r > .41$, $p < .01$; [a]$N = 37$.

tionships were obtained between percent fat and endomorphy, which is to be expected. Similar studies on women students by Slaughter and Lohman (1976), Slaughter, Lohman, and Boileau (1977), Bale (1981), and on children by Hunt and Barton (1959) and Slaughter and Lohman (1977) also indicate a close relationship between endomorphy and percent fat.

In the men students, both lean body weight and mesomorphy were moderately related with the index of total strength; there were also moderate relationships between mesomorphy and right grip, leg lift and back strength, and between lean body weight and all the strength measures. Though there were moderate relationships between total strength and mesomorphy and between total strength and lean body weight in the women students, the somatotype and body composition variables showed little relationship with the individual strength measures. The findings suggest that in men, the higher their muscularity rating (whether expressed as mesomorphy or lean body weight) the greater their overall strength. This relationship is also found in women students but to a lesser degree. However, care must be taken not to overemphasize the relationship between morphology and strength on the basis of the above findings, as many of the individual strength measures showed little relationship with somatotype and body composition, especially in the women students.

Just as ratings for mesomorphy or assessment of lean body weight are general indicators of the muscular content of the body, the total strength index can be considered a general indicator of overall static strength. The individual strength measures which contribute to this strength index each indicate strength in one region of the body. Thus more precise measures of the musculature of these regions would be required to obtain a clearer relationship between the individual strength measures and physique. In a separate study on 51 of the sport science students, the relationship between forearm circumference and grip strength was therefore examined. Product-moment correlations of .54 and .43 between the men's right forearm circumference and right grip, and left forearm circumference and left grip were obtained. Similar correlations of .58 and .52 were obtained between the women's forearm circumferences and their grip strengths. It would be interesting to examine the relationships between other variables such as biceps circumference and arm strength or between leg volume and leg lift.

Table 5 presents the means and standard deviations of the physique, body composition, and strength variables of the men sport science students when classified according to their sport specialties. Those students specializing in sports such as sailing, rowing, cricket, and cycling were so few that they were omitted from this part of the analysis. Even so, the number of subjects in some of the other sport groups was so small that only tentative comparisons between sports can be made. Of the nine men's sport groups, the middle- and long-distance runners were lightest. Their mean weight of 66.5 kg was significantly lighter than the mean weights of the sprinters and jumpers, rugby players, basketball players, and weight lifters. The weight lifters were heaviest (mean weight of 79.1 kg), yet were the shortest of the sportsmen, apart from the rugby players.

The middle- and long-distance runners were significantly lower in mesomorphy than the other sport groups, and they had significantly higher ectomorphy

Table 5. Means and standard deviations of the physique, body composition, and strength measures of the men's sport groups

Sport group		1 Height (cm)	2 Weight (kg)	3 Endo-morphy	4 Meso-morphy	5 Ecto-morphy	6 Ponderal index	7 Total skinfold (mm)	8 Density
Soccer	M	181.7	76.9	2.7	4.7	3.2	13.0	36.3	1.0720
n = 11	SD	5.2	7.0	0.6	0.6	0.8	0.4	8.0	.0049
Rugby	M	177.1	76.7	2.3	5.4	2.4	12.7	34.8	1.0715
n = 12	SD	6.0	8.0	0.7	0.7	0.7	0.3	9.6	.0067
Rackets	M	178.5	71.1	2.5	4.7	3.6	13.2	33.9	1.0731
n = 10	SD	4.8	5.1	0.7	0.7	0.8	0.3	11.1	.0081
Swimming	M	179.9	71.3	2.1	4.9	3.3	13.1	28.8	1.0778
n = 7	SD	5.8	5.6	1.0	0.7	1.0	0.4	8.5	.0060
Middle &	M	178.9	66.5	2.2	3.6	4.4	13.4	24.5	1.0803
long distance	SD	8.5	7.7	0.5	0.8	1.0	0.4	3.3	.0033
n = 9									
Sprinters &	M	183.4	75.2	2.1	4.8	3.3	13.1	26.7	1.0782
jumpers	SD	3.6	4.3	0.5	0.4	0.9	0.3	6.4	.0054
n = 6									
Golfers	M	178.4	70.1	2.0	4.5	3.4	13.1	31.6	1.0769
n = 5	SD	5.0	7.0	1.0	0.0	0.9	0.4	14.9	.0092
Basketball	M	184.2	76.9	2.5	4.7	3.6	13.1	29.9	1.0752
n = 5	SD	7.4	7.6	0.8	0.4	0.8	0.3	7.0	.0053
Weight lifters	M	177.7	79.1	2.3	6.3	1.5	12.4	32.3	1.0735
n = 4	SD	5.2	9.3	0.5	1.3	0.6	0.4	8.1	00.57
Total	M	179.7	73.6	2.3	4.8	3.2	13.0	31.5	1.0750
n = 69	SD	6.0	7.6	0.7	0.9	1.1	0.4	9.3	.0066
F value		1.2	2.7*	0.9	7.0**	6.2**	5.2**	1.7	2.2*

$*p < .05;\ **p < .01.$

ratings and ponderal indices than the soccer players, rugby players, swimmers, sprinters and jumpers, and weight lifters. They also had the highest body densities and the lowest fat measurements and lean body weights. The mean somatotype of these endurance runners is less mesomorphic and more ectomorphic than the mean somatotypes of the other sport groups, particularly the weight lifters. There were significant differences in strength between the middle- and long-distance runners and the other sport groups. As expected, the weight lifters had the highest strength measures, followed by the basketball, soccer, and rugby players.

Fewer significant differences were found among the five groups of women sport science students (Table 6). The sports acrobats were the shortest and also the lightest (apart from the middle- and long-distance runners), but they had the highest mesomorphy ratings. The middle- and long-distance women runners had significantly higher mean ectomorphy ratings and ponderal indices than the other four sport groups. Like their male counterparts, the women middle- and long-distance runners have somatotypes closer to the ectomorphy

Table 5. Means and standard deviations of the physique, body composition, and strength measures of the men's sport groups

9	10	11	12	13	14	15	16	17	18
% fat	Absolute fat (kg)	Lean body weight (kg)	Right grip (kg)	Left grip (kg)	Right arm (kg)	Left arm (kg)	Back strength (kg)	Leg strength (kg)	Overall strength (kg)
11.8	9.1	67.8	51.6	47.2	26.8	23.6	147.5	190.6	487.4
2.1	2.1	5.8	6.3	5.9	4.2	3.9	13.0	19.8	25.9
12.0	9.3	67.4	52.4	47.5	27.2	25.6	140.5	198.1	486.0
2.9	3.1	5.8	7.2	5.6	3.9	4.5	22.4	25.0	50.7
11.3	8.2	62.9	48.0	42.6	22.3	19.8	122.3	183.1	438.2
3.5	3.1	2.8	5.7	5.1	4.8	3.5	22.5	41.9	49.9
9.3	6.7	64.6	52.5	48.4	22.8	21.2	135.7	186.3	465.4
2.6	2.1	4.4	8.5	7.1	2.9	1.8	20.2	20.2	23.0
8.2	5.5	61.0	45.7	40.6	20.7	21.1	115.0	160.2	402.2
1.4	1.3	6.7	7.9	6.8	4.6	4.7	19.8	24.7	53.3
9.1	6.9	68.4	53.4	49.1	27.6	28.4	140.8	186.8	485.9
2.3	1.8	3.7	6.5	4.1	5.6	3.9	27.7	35.7	68.1
9.7	6.9	63.1	52.5	52.7	25.0	24.0	132.4	163.3	449.9
3.9	3.4	4.8	3.6	7.8	4.5	8.3	14.0	17.2	36.4
10.4	8.1	68.8	55.3	50.9	26.8	25.5	147.4	202.2	508.1
2.3	2.5	6.0	7.2	7.9	2.8	3.9	14.6	30.3	58.2
11.1	9.0	70.2	63.1	55.6	31.3	28.3	157.7	219.9	555.8
2.5	3.1	6.4	7.6	9.6	4.7	1.7	50.8	41.5	93.0
10.5	7.9	65.7	51.8	47.2	25.3	23.7	136.2	187.1	469.9
2.9	2.7	5.8	7.6	7.3	5.0	4.9	24.8	31.6	63.6
2.2*	2.3*	2.6*	2.8*	3.4**	3.9**	3.5**	2.2*	2.3*	3.9**

axis of the somatochart with the lowest endomorphy and mesomorphy ratings.

The above findings support the view that a high mesomorphy rating and lean body weight are important for sports requiring strength and power (such as weight-lifting, contact sports, or gymnastics), whereas a slim physique is best suited to sports requiring steady-state activity (such as long-distance running) and high cardiorespiratory efficiency. Only tentative comparisons can be made between this study and other studies because different methodologies have been used, the athletes may have come from different ethnic groups, and the level of performance varies. However, the physiques and body composition of these sport science students are similar to those of top class men and women athletes reported by other researchers (Fleck, 1983; Wilmore, 1983). Many of the sport science students have represented their country at junior level in their respective sports (for example, at the World Student Games), and they are considered potential international performers. Some of the swimmers, racket players, athletes, and distance runners have already achieved full international honors.

Table 6. Means and standard deviations of the physique, body composition, and strength measures of the women's sport groups

		1	2	3	4	5	6	7	8
Sport group		Height (cm)	Weight (kg)	Endo-morphy	Meso-morphy	Ecto-morphy	Ponderal index	Total skinfold (mm)	Density
Hockey	M	162.7	59.4	3.8	4.4	2.6	12.7	57.3	1.0510
n = 11	SD	4.6	7.5	0.8	0.8	1.1	0.5	16.6	.0061
Racket sports	M	166.9	61.9	3.6	3.9	2.6	12.9	55.1	1.0505
n = 9	SD	5.1	6.6	0.9	0.4	1.0	0.3	14.4	.0053
Sports	M	158.9	56.1	3.4	4.9	2.1	12.6	45.4	1.0551
acrobatics	SD	4.3	6.5	0.5	1.0	1.0	0.4	5.7	.0034
n = 4									
Middle &	M	166.9	55.0	2.7	3.6	4.1	13.3	39.8	1.0580
long distance	SD	2.6	5.8	1.0	0.9	1.0	0.4	11.9	.0070
n = 5									
Basketball	M	169.2	66.8	3.9	4.3	2.5	12.7	55.7	1.0514
n = 5	SD	4.9	8.2	0.5	0.4	0.9	0.3	16.1	.0073
Total	M	164.9	60.1	3.5	4.2	2.8	12.8	52.5	1.0524
n = 34	SD	5.3	7.6	0.9	0.8	1.1	0.5	15.1	.0063
F value		4.2**	2.3	1.9	2.3	2.7*	2.9*	1.6	1.7

$*p < .05; **p < .01.$

References

Bale, P. (1979). The relationship between physique and basic motor performance in a group of female P.E. students. *Carnegie Research Papers*, **1**, 26-32.

Bale, P. (1980). The relationship of physique and body composition to strength in groups of female physical education students. *British Journal of Sports Medicine*, **14**(4), 193-198.

Bale, P. (1981). Body composition and somatotype characteristics of sportswomen. In J. Borms, M. Hebbelinck, & A. Venerando (Eds.), *The female athlete: Vol. 15. Medicine and sport* (pp. 157-167). Basel: Karger.

Carter, J.E.L. (1970). The somatotypes of athletes, a review. *Human Biology*, **42**, 535-569.

Carter, J.E.L. (1981). Somatotypes of female athletes. In J. Borms, M. Hebbelinck, & A. Venerando (Eds.), *The female athlete: Vol. 15. Medicine and sport* (pp. 85-116). Basel: Karger.

Damon, A., & Goldman, R.F. (1964). Predicting fat from body measurements: densitometric validation of ten anthropometric equations. *Human Biology*, **36**, 32-44.

de Garay, A.L., Levine, L., & Carter, J.E.L. (1974). *Genetic and anthropological studies of Olympic athletes*. New York: Academic Press.

Desipre, M., Van Reenen, O.R., Gelderblom, I., & Barnard, J. (1976). Validity and reliability of determining fat free mass, body fat and relative fat in young athletes. *South African Journal for Research in Sport, P.E. and Recreation*, **2**(1), 42-62.

Durnin, J., & Rahaman, M. (1967). The assessment of the amount of fat in the human body from measurement of skinfold thickness. *British Journal of Nutrition*, **21**, 681-689.

Durnin, J., & Womersley, J. (1974). Body fat assessed from total body density and its estimation from skinfold thickness: Measurements on 481 men and women aged 16 to 72 years. *British Journal of Nutrition*, **32**, 681-689.

Table 6. Means and standard deviations of the physique, body composition, and strength measures of the women's sport groups

9	10	11	12	13	14	15	16	17	18
% fat	Absolute fat (kg)	Lean body weight (kg)	Right grip (kg)	Left grip (kg)	Right arm (kg)	Left arm (kg)	Back strength (kg)	Leg strength (kg)	Overall strength (kg)
21.0	12.6	46.8	34.9	32.2	16.0	14.0	93.6	129.0	323.2
2.8	2.0	2.0	7.1	5.8	2.6	2.6	13.7	21.8	43.0
21.2	13.2	48.7	36.2	31.7	16.8	16.1	87.7	107.7	296.4
2.4	2.5	4.3	6.2	6.2	2.4	3.6	15.4	7.5	26.6
19.2	10.8	45.3	37.0	32.5	16.8	16.1	108.5	122.7	333.6
1.5	1.8	4.9	6.1	4.3	1.3	2.9	17.7	23.4	51.8
17.9	10.0	45.0	33.3	30.1	17.4	15.6	97.2	130.1	323.8
3.1	2.5	3.5	4.0	5.1	3.9	2.0	18.0	18.4	39.0
20.8	14.1	52.8	38.7	32.9	18.1	17.3	103.3	124.8	341.1
3.2	3.3	5.4	6.4	5.5	2.5	2.5	13.7	16.9	33.5
20.4	12.4	47.7	35.8	31.9	16.8	15.5	96.0	122.3	320.0
2.9	2.9	5.0	6.1	5.4	2.6	2.9	15.8	19.2	39.3
1.7	2.0	2.3	0.6	0.2	0.6	1.4	1.6	1.9	1.2

Fleck, S.J. (1983). Body composition of elite American athletes. *American Journal of Sports Medicine,* **11**(6), 398-402.

Forsyth, H., & Sinning, W.E. (1973). The estimation of L.B.W. in male athletes. *Medicine and Science in Sports,* **5**, 174-180.

Harsha, D., Fredrichs, R., & Berenson, G. (1978). A simple and complete densitometric technique for under-water weighing. *Journal of Sports Medicine and Physical Fitness,* **18**, 253-262.

Heath, B.H., & Carter, J.E.L. (1967). A modified somatotype method. *American Journal of Physical Anthropology,* **27**, 57-74.

Heyward, V., & McCreary, L. (1977). Analysis of the static strength and relative endurance of women athletes. *Research Quarterly,* **48**(4), 703-710.

Hunt, E., & Barton, W.H. (1959). The inconstancy of physique in adolescent boys and other limitations of somatotyping. *American Journal of Physical Anthropology,* **17**, 27-35.

Jackson, A., Pollock, M., & Ward, A. (1980). Generalized equations for predicting body density of women. *Medicine and Science in Sport,* **12**(3), 175-182.

Katch, F.I., & Katch, V.L. (1980). Measurement and prediction errors in body composition assessment and the search for the perfect prediction equation. *Research Quarterly,* **51**(1), 249-260.

Katch, F.I., & Michael, E. (1968). Prediction of body density from skinfolds and girth measurements of college women. *Journal of Applied Physiology,* **25**, 92-94.

Montoye, H.J., & Lamphiear, D.E. (1977). Grip and arm strength in males and females, age 10 to 69. *Research Quarterly,* **48**(1), 109-120.

Parnell, R.W. (1958). *Behaviour and physique.* London: Arnold.

Powers, S.K., & Walker, R. (1982). Physiological and anatomical characteristics of outstanding female tennis players. *Research Quarterly,* **53**(2), 172-175.

Sinning, W.E. (1980). Use and misuse of anthropometric estimates of body composition. *Journal of Physical Education and Recreation,* **51**(2), 43-45.

Siri, W.E. (1956). The gross composition of the body. In J.E. Lawrence & C.A. Tobias (Eds.), *Advances in biological and medical physics*. New York: Academic Press.

Slaughter, M.H., & Lohman, T.G. (1976). Relationship of body composition to somatotype. *American Journal of Physical Anthropology, 44*, 237-244.

Slaughter, M.H., & Lohman, T.G. (1977). Relationship of body composition to somatotype in boys aged 7 to 12 years. *Research Quarterly, 48*(4), 750-758.

Slaughter, M.H., Lohman, T.G., & Boileau, R. (1977). Relationship of Heath and Carter's second component to lean body mass and height in college women. *Research Quarterly, 48*(4), 759-768.

Sloan, A.W. (1967). Estimation of body fat in young men. *Journal of Applied Physiology, 23*, 311-315.

Sloan, A.W., Burt, J.J., & Blyth, C.S. (1962). Estimation of body fat in young women. *Journal of Applied Physiology, 17*, 967-970.

Wilmore, J.H. (1983). Body composition in sport and exercise: Directions for future research. *Medicine and Science in Sports, 15*(1), 21-31.

Wilmore, J.H., & Behnke, A.R. (1968). Predictability of lean body weight through anthropometric assessment in college men. *Journal of Applied Physiology, 25*, 349-355.

Wilmore, J.H., & Behnke, A.R. (1970). An anthropometric estimation of body density and lean body weight in young women. *American Journal of Clinical Nutrition, 23*(3), 267-274.

Wilmore, J.H., Vodak, P.A., Parr, R.B., Girandola, R.H., & Billing, J.E. (1980). Further simplification of a method for determination of residual volume. *Medicine and Science in Sports, 12*(3), 216-218.

PART V

A Techniques Perspective

"Fine-tuning the instrument" is the thread linking the papers in Part V. Whereas the previous papers in this volume present the results of kinanthropometric research, the reports in Part V are concerned with how our research techniques can be improved. We accept that progress is based on better anthropometry, better technology, better theoretical work, and on more reasoned argument. Some of each of these is present in this section.

Five of the nine papers have body composition as their major or minor focus. The keynote paper by William Ross and his colleagues considers at length the difficulties, both theoretical and practical, which inhibit the usefulness of current methods of body composition assessment. Their *O-Scale technique*, which avoids the pitfalls of "percent fat," is the latest word in this area, but probably will not be the last. The other papers presenting modified techniques for body composition assessment are those by Drinkwater and his colleagues, Sinning and Hackney, Svoboda and Query, and Donnelly and Sintek.

Van der Walt and his co-workers have done extensive practical work and insightful theoretical work in the development of their paper on muscle volume. Their paper is important both for the data they present and for the novel methodology they propose. Day's paper on upper-limb measurement techniques can lead to greater precision in basic anthropometry. The final two papers, by Woltring and Baumrind, were selected from among those included in the symposium on the application of biostereometrics to sport. This symposium was chaired by Dr. Bhim Savara, a member of the Scientific Program Commission for the OSC. The future of kinanthropometry is clearly laid before us by these important papers.

19

Alternatives for the Conventional Methods of Human Body Composition and Physique Assessment

William D. Ross
SIMON FRASER UNIVERSITY
BURNABY, BRITISH COLUMBIA, CANADA

*Richard Ward, Alan D. Martin,
and Donald T. Drinkwater*
SIMON FRASER UNIVERSITY
BURNABY, BRITISH COLUMBIA, CANADA

Otto G. Eiben
EOTVOS LORAND UNIVERSITY
BUDAPEST, HUNGARY

Jan P. Clarys
VRIJE UNIVERSITEIT BRUSSEL
BRUSSELS, BELGIUM

In a mathematical sense, the word *elegant* applied to problem-solving means "neat, ingeniously simple, and effective." In all of science, the use of the ratio of mass and volume in Archimedes' principle is one of the most elegant solutions to a problem ever proposed. The purpose of this paper, thus, is to discuss basic issues regarding the use of ratios in human biology and to propose alternate solutions for the assessment of body composition and the appraisal of proportionality characteristics.

We share with other kinanthropometrists a debt of gratitude to our Belgian colleague, Dr. J.P. Clarys, Vrije Universiteit Brussel, who directed the basic work in body composition which has spawned new methods and approaches. We also acknowledge Drs. J. Borms and M. Hebbelinck of the same university for their contribution in the assembly of data on bodybuilders and the support of President Ben Wieder of the International Federation of Bodybuilders. We are continually indebted to Drs. D.A. Bailey and R.L. Mirwald for access to their longitudinal growth data and the large data base from their YMCA Lifestyle Inventory and Fitness Evaluation Project. We appreciate, too, our colleagues in Canada and Hungary who have unstintingly helped in data assembly and resolution. Large, reliable data bases provide for many experiments and tests necessary for developing new models and approaches. Recent basic support for the study of human and animal growth at Simon Fraser University was from NSERC grant A9402; support for completion of the O-Scale System was from Fitness Canada grants in 1983 and 1984.

203

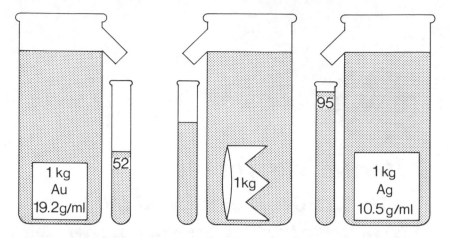

Figure 1. Volume and density of gold, silver, and the alloy of the crown.

Archimedes (278 to 212 B.C.) allegedly discovered the basic principle while sitting in a bathtub. At the time he was aware of the rumor that an artisan had substituted silver for part of the weight of gold he had been given by the King of Syracuse to fashion a golden wreath or crown. As he was immersing in the tub, he noted that the amount of water spilling over the sides was related to the degree he immersed. This was the trigger for sudden enlightenment about the principle of buoyancy, and, as legend has it, he sprang naked from the bath and ran into the street proclaiming, "Eureka, I have found it."

From the understanding of buoyancy and density, he was able to expose the artisan as a fraud. Gold, being heavier than silver for a given volume, has a density of 19.2 g/ml. A kilogram of gold would have a volume of 1000/19.2 or 52.1 ml, whereas a kilogram of silver, the baser metal with a density of 10.5 g/ml would have a volume of 1000/10.5 or 95.2 g/ml. As illustrated in Figure 1, the crown displaced more water than possible for pure gold. Therefore, it must have been alloyed, at a profit, by some baser metal. From the density of an object and known densities of two constituent parts, one may calculate the proportion of each which constitutes the whole. This is illustrated graphically for the case of the golden crown in Figure 2.

Body Composition

Density and Hydrostatic Weighing

Since the 1940s, the elegant Archimedes' principle has been applied to the study of human body composition. The assessment of whole body density is done by underwater weighing or displacement with a correction for the buoyant force of entrapped air in the lungs and viscera. The determination of whole body density, particularly by underwater weighing, has its own imprinting of elegance. When one knows the weight of the body in air, its weight in water, and the density of the water, it is possible to determine whole body density

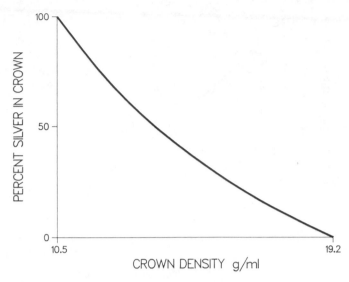

Figure 2. Variation of silver content with crown density.

with appropriate correction for lung and visceral gases. The technique is fairly standard. However, it has rather stringent technical requirements, particularly if one is to assume underwater weight is "minimal" and all entrapped air is exhaled except for the residual volume that is used as a correction factor.

The excellent paper by Katch (1969) is instructive in this regard. He found learning effects in achieving minimal weight in as many as 15 trials. Moreover, from our own observations we have yet to find a subject who did not breath-hold, at least to the point of having ample air entrapment to blow clear the nasal passage upon resurfacing from submergence in the tank. For argument's sake, we might concede that it is technically possible to obtain accurate measures of whole body density, if not by a full-exhalation method, at least by breath-holding and correcting for the actual volume of entrapped air (Ross & Marfell-Jones, 1982). The so-called visceral gas is really indeterminable and subtraction of an estimated value such as $115(170.18/h)^3$, which takes into account body size, is as good as any method.

Percent Fat by Densitometry

In order for the percent fat formulae used in densitometric equations to work, one must make two basic assumptions: (a) The human body has two compartments, *fat* and *nonfat*, and (b) each of these has densities which are known constants. One of the most popular paradigms, that of Siri (1961), assumes fat to have a density of 0.90 g/ml and nonfat a density of 1.10 g/ml. As shown in Figure 3, the Archimedes' model relates whole body densities to the scale 0.90 g/ml for 100% fat and to 1.10 g/ml for 0% fat. The classification shown in Table 1 is entirely arbitrary. By usage of this or other such classifications, percent fat estimates have received some operational validity in fitness appraisal and guidance. However, the density assumptions have never been validated

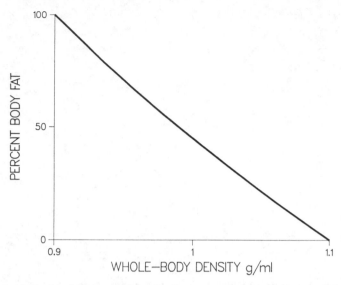

Figure 3. Variation in percent fat with body density.

Table 1. Percent fat from body density for young adults

Density (g/ml)	% fat	Classification
1.090	4.1	Very lean male
1.080	8.3	Lean male
1.070	12.6	Lean female Average male
1.060	17.0	Desirable female Fat male
1.050	21.4	Average female Very fat male
1.040	26.0	Plump female Obese male
1.030	30.6	Fat female
1.020	35.5	Very fat female
1.010	40.0	Obese female

by direct evidence, nor has it been established except in a very gross way that any classification is optimal for any individual.

The Implication

The acceptance of the densitometric assumptions and the extrapolation to percent fat leads to the use of hydrostatic weighing as a universal method for validating skinfold caliper formulae and all other methods. Wilmore (1983) refers to the densitometric method as the "gold standard."

At this point in our exposé, we like to tell the story about the man who stopped every morning at the jeweler's store to adjust his watch to the time displayed on the chronometer in the window.

One day the jeweler asked, "Why do you need such accurate time?" He replied, "I have a very important job. I am the man who blows the noon whistle." "Oh my goodness," said the jeweler, "I calibrate my chronometer by the noon whistle."

Apart from the anatomical approach to the assessment of human body composition by Matiegka (1921) and a few other propositions, the preponderance of body composition methodology is vulnerable to the extent that the underlying densitometric assumptions are in error.

The Issues

Our group has always had great trepidation in assuming constancy of any biological entity. The division of the body into a fat compartment (presumably ether extractible lipid) with a constant density might be challenged because the extract includes phospholipids, glycolipids, sterols, and glycerides wherever they are located and with whatever structures they are associated. However, for the sake of argument, we shall leave the constancy of the fat density as a moot question. We feel this is largely irrelevant because our advocacy is for an *anatomical* and biologically meaningful fractionation of the body and not a *chemical* fractionation. In the chemical fat-nonfat dichotomy, the constancy of the nonfat compartment implicitly requires that two conditions must be met: (a) The constituent tissues in the nonfat compartment are present in fixed proportions, and (b) the densities of the constituent tissues each are constant and known.

The Evidence

There is *no* direct human anatomical evidence that either of the above two conditions is met. Cadaver evidence from an anatomical division into dissectible *adipose* and *adipose tissue free* masses is not at all supportive (Martin, 1984). First, the proportions of constituent tissues of the adipose tissue free mass of 6 male and 7 female unembalmed cadavers were *not fixed*; muscle ranged from 41.9% to 59.4%, bone from 16.3% to 25.7%, and the residual or remainder from 24.0% to 32.4%. Second, the density of the dissectible muscle of the 13 unembalmed cadavers was shown to be relatively stable with a coefficient of variation of approximately 1%, about a mean density of 1.05 g/ml. Bone, however, appears to have an even greater variability as shown in the summary of 25 embalmed and unembalmed cadavers in Table 2 (Martin, 1984).

There was no significant sex difference in skeletal density in the cadaver sample, although there was an age-associated decrement of about 0.02 g/ml per decade. When projected over the entire age range, including children, we should expect even greater deviation from the assumed constancy.

The differing proportions and density of constituent tissues in the adipose tissue free mass strongly mitigate against the essential densitometric assumptions of constancy of the nonfat compartment. The cadaver evidence helps explain nonsensical extrapolations such as "negative body fat" from densitometric findings in excess of 1.10 g/ml (Adams, Mottola, Bagnall, & McFadden, 1982) or values of 1, 2, or 3% where presumably the amount predicted is less than

Table 2. Mean densities (at 20° C) of selected bones from the dissection of 25 cadavers

Bone	Density	
	M	SD
Pelvis	1.164	0.037
Tibia	1.242	0.055
Humerus	1.262	0.054
Femur	1.267	0.053
Clavicle	1.315	0.058
Radius	1.353	0.070
Ulna	1.395	0.078
Cranium	1.403	0.061
Mandible	1.570	0.100
Whole skeleton	1.236	0.039

the amount of essential lipid to support life processes. It is reasonable to surmise that variance at the other end of the density continuum is also operative, although errors are not signalled by negative values. Instead of the single curve linking 0.90 to 1.10 g/ml required for the two-compartment model, if the nonfat compartment varies to the extent of the adipose tissue free mass, we are likely to have a family of curves as shown in Figure 4.

Demise of the Density Criterion

The measurement of height, weight, and body volume provides basic parameters for human body composition. Whole body density is an excellent indicator for change in compositional status, especially when used in conjunction with comprehensive anthropometry. However, using it as a universal criterion for individual prediction of percent fat is untenable.

Quetelet and the Weight-Height Indices

The Belgian astronomer and mathematician Lambert Adolphe Quetelet (1796-1874) might be identified as the father of both kinanthropometry and physical anthropology. He discovered that the error distribution which worked so well in describing astronomical measures was also a reasonably good model for the empirical distribution of anthropometric and other measures on humans. One hundred and fifty years ago he recorded chest girths of Scottish soldiers, the stature of French army draftees, and other such measures, and found these distributed around the "average" in a predictable fashion. He studied bodily proportions in classical and Renaissance art and compared findings to cross-sectional data on Belgians of his time. He conducted longitudinal studies on two of his daughters and was an avid student of ethnic and genetic differences and the variety of the human species.

As pointed out by Ross, Drinkwater, Bailey, Marshall, and Leahy (1980), Quetelet was not without his critics. Over 100 years after his death, Hogben (1957) decried the "Quetelet mystique," or that alleged influence whereby

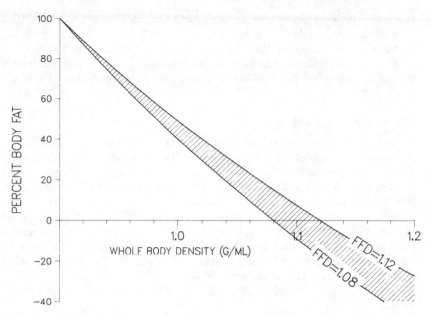

Figure 4. Siri's equation for the fat-free densities in the range 1.08 to 1.12 g/ml.

investigators persist in regarding the normal probability curve as the population archetype. Surely, if there is reason for censure, it is not with Quetelet, but with the unthinking investigators who see all arrays of data as being normally distributed and reflexively apply F or t-tests. In this, one is tempted to carry the reflex analogy further and to suggest the data stimulus and response is linked by an interneuron which bypasses the higher centers in the brain. It is more appropriate to note that Quetelet was at least 150 years ahead of his time, and much of what he wrote is contemporary and ought to be appreciated as part of our scientific legacy (Jelliffe & Jelliffe, 1979).

In recent years, largely because epidemiologists have had mostly only height and weight data on large cross-sectional samples, a parade of investigators has used the Quetelet index, that is, weight divided by height squared (w/h²), in reporting and interpreting data, thus inferring obesity, adiposity, or relative fatness. In one of his early books, Quetelet (1836) developed the concept of a universal prototype, "the average man," about whom measurements are dispersed according to the normal probability curve. In the same book, he proposed his time-honored Quetelet index, or the quotient of weight when divided by height squared (w/h²), based on calculations that best described the relationship between weight and height of individuals. Dimensionally, this implies a systematic trend to linearity with increasing size and suggests the human design is toward long-lankiness and short-squatness.

Rationale for Weight-Height Indices

The contemporary use of weight-height ratios is mainly by health scientists studying epidemiological problems, usually from cross-sectional sampling surveys involving relatively few anthropometric measures. The premise in using the indices is that body weight corrected for stature is correlated with obesity or adiposity. In accepting the premise, at least for discussion, the problem then becomes to determine the index which is maximally correlated with weight and minimally correlated with height, that is, the index least biased by height in all populations.

The Supporting Evidence

In population studies there is considerable support for the use of the Quetelet index (w/h^2) as opposed to other indices such as (w/h) and (w/h^3). In samples from various populations (w/h^2) has had low correlations with height as shown by Billewicz, Kemsley, and Thompson (1962), Visweswara and Singh (1970), Khosla and Lowe (1967), Keys, Fidanza, Karvonen, Kimura, and Taylor (1972), Cerovska, Petrasek, Hajino, and Kaucka (1977), Babu and Chuttani (1979), and Frisancho and Flegel (1982). Furthermore, the Quetelet index has been positively correlated with skinfold caliper thicknesses and, in some instances, with underwater weight, as reported by Florey (1970), Benn (1971), Keys et al. (1972), Goldbourt and Medalie (1974), Cerovska et al. (1977), Roche, Siervogel, Chumlea, and Webb (1981), and Frisancho and Flegel (1982).

The Quetelet index is not without criticism. Lee, Kolonel, and Hinds (1981) contended that it often had a substantial correlation with height and that a better ratio is that proposed by Benn (1971), where weight is divided by height to the actual dimensional exponent obtained in the allometric equation of height predicting weight (w/h^p). This has elicited debate with Knowler and Garrow (1982) who challenge the need for the "independence of height criterion" and state that the Benn index is a specific solution not substantiated in real data. Lee, Kolonel, and Hinds (1982) and Lee and Kolonel (1983) insisted that height-weight ratios other than those based on sample dimensions could be misleading; notwithstanding, Garn and Pesick (1982), in summary analysis of 58,468 individuals in 10 samples from four surveys, concluded it was not evident that the population specific p in the Benn index conveys a material advantage in nutritional surveys.

In a factorial analysis of 961 obese patients, Colliver, Stuart, and Arthur (1983) concluded that all of the two-factor weight-height indices were measuring essentially the same thing and that "relative adiposity," however defined, should be virtually identical. They further concluded that it should make little difference which of the six commonly used indices is used in the clinical appraisal of obese adults. Which of the proponents in the debate are right? We suggest none.

Growth and maturation, as well as environmental factors, mitigate against biological constancy of tissue masses. Malina, Meleski, and Shoup (1982) showed that systematic training effects are detectable in children, although the dimensional implications have yet to be thoroughly explored. A principal components analysis from six American longitudinal growth studies by Cronk et al. (1982) suggested that a significant proportion of adult fatness was not predictable in childhood from the weight for height ratio. Billewicz, Thompson, and Fellowes (1983) also expressed trepidation in the interpretation of weight for height ratios when there is a compounding of pubertal growth effects. With interpretive cautions, Rolland-Cachera et al. (1982) have projected norms for the Quetelet index from longitudinal data.

Much confusion about human dimensionality could be reduced if referees for journals would insist that the actual dimensional relationships and some indication of the predictive power of the regression equation be reported. Allometric *b*-values departing from theoretical values should signal inappropriateness of the assumptions in the scaling unless the purpose in scaling is to study the departures as we discuss in the section on conventionalist solutions.

The Indictment of Weight-Height Indices

Anatomical evidence from cadaver studies clearly shows what all human biologists who understand biological variation must know intuitively: Body weight is not a simple entity but is composed of different proportions of tissues with varying densities. While one might make some gross generalizations about a sample and make group comparisons, it is quite another matter to make individual predictions.

It is easy to demonstrate gross errors in using weight-height ratios as indices of adiposity. We can show this in a sample of 221 male and 177 female university students from our data base and a sample of 66 male participants in the 1981 International Federation of Body Builders' World Championships in Cairo, Egypt. As shown in Table 3, the Quetelet index was not correlated with height for the males ($r = -0.08$) or for the females ($r = -0.01$) exactly meeting the height dissociation criterion. The index was, however, significantly correlated with height in the body builder sample ($r = 0.41$).

Using the procedures of proponents of the Quetelet index, we compared the index with the sum of skinfolds, as graphically displayed in Figure 5. For our sum of skinfolds we used the median of three or mean of two skinfold measures obtained at the triceps, subscapular, supraspinale (suprailiac), abdominal, front thigh, and medial calf sites. As summarized in Table 3, the Quetelet index

Table 3. Correlation of height with weight-height ratios, the sum of six skinfolds (S6SF), and dimension of height with weight for Canadian males and females and male bodybuilders

Variable	HT	WT	W/HT	W/HT²	W/HT³
	Males (n = 221, b value = 1.78)				
HT	—	0.54	0.27	-0.08	-0.40
WT	0.54	—	0.95	0.79	0.54
S6SF	-0.11	0.45	0.55	0.62	0.60
	Females (n = 177, b value = 1.97)				
HT	—	0.64	0.38	-0.01	-0.39
WT	0.64	—	0.95	0.76	0.44
S6SF	0.06	0.45	0.52	0.54	0.48
	Bodybuilders (n = 66, b value = 2.75)				
HT	—	0.85	0.72	0.41	-0.14
WT	0.85	—	0.98	0.82	0.40
S6SF	0.15	0.15	0.15	0.18	0.05

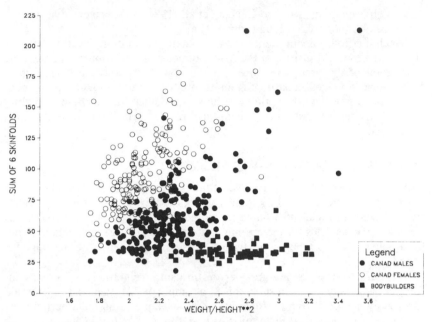

Figure 5. Quetelet's index predicting the sum of six skinfolds.

(w/h²) was significantly correlated with the sum of the skinfolds for the males ($r = 0.62$) and for females ($r = 0.54$). It was not correlated in the body builder sample ($r = 0.18$). The correlations for the male and female samples were somewhat lower than might be anticipated, perhaps because the university sample represented relatively lean subjects. Clearly, prediction of an adiposity level from a correlation of 0.62 is far from adequate for individual prediction, because the predictive index is only 22% better than chance. In some samples where the subjects are under training or skinfold thicknesses are minimal (as demonstrated in the body builder sample), there is no appreciable relationship of the weight-height ratio with skinfold totals.

Obesity Screening: The O-Scale System

In his book on understanding science, James B. Conant (1947, p. 90) referred to "the principle that a conceptual scheme is never discarded because of a few stubborn facts with which it cannot be reconciled; a concept is either modified or replaced by a better concept, never abandoned with nothing better left to take its place."

For simple identification of differing levels of adiposity and scaling to some standard expectancy, we feel the use of a weight-height ratio, by itself, is indefensible. We accept that skinfold calipers, at best, must regionally sample the body, and they must be size-dissociated if they are to be used as an obesity indicator. We also feel that a measure of ponderosity (weight for height) is useful and, in this, we prefer to view the body in terms of departures from geometric similarity, accepting that tall individuals will tend to be less ponderous than short individuals, because proportionally, they tend to have longer legs, shorter arms and trunks, narrower shoulders and smaller chest measures, particularly chest depth.

The O-Scale system proposed by Ross and Ward (1985) essentially consists of percentile-transformed, geometrically size-adjusted skinfold totals that are displayed concurrently with proportional body weight on standard nine, or stanine scales, for males and females yearly from age 6 to 17, 18, and 19, and in 5-year increments thereafter until age 70. The data base for the present system consists of about 24,000 subjects from our own studies and that of the YMCA Lifestyle Inventory and Fitness Evaluation program by Drs. D.A. Bailey and R.L. Mirwald of the University of Saskatchewan. Possibly the sample has a disposition toward active health. Further augmentation and comparison with geographically stratified, more randomly selected subjects are anticipated when the Canada Fitness Survey data are resolved. Even should there be systematic differences, one may contend, as does Tanner (1976), that norms based on a sample of "best off" subjects may be the most appropriate for assessing and monitoring health status.

It is beyond the scope of this paper to discuss measurement technique, certification procedures, and access to microcomputer software and services. However, a single printout of the data assembled on the lightweight body building champion from the 1981 IFBB championships will serve to illustrate the accommodation of extreme physiques in a system which scales all data to a common stature and rates on a 9-point *leanness to obesity* scale for the skinfolds and *linearity to ponderosity* scale for the proportional body weight. As shown in Figure 6, the subject's size-adjusted skinfold value for men his own age gave him a 2 rating, placing him between the 4th and 11th percentiles. However, his pronounced muscularity raised his porportional body weight to a 9 because he was proportionally much heavier than most men of his age. All of the traditional indices would have classified him as "obese." Even in subjects with less extreme physiques, it is advantageous to separate the assessment of adiposity and ponderosity in morphometric analyses.

More on Proportionality

Many other traditional anthropometric ratios have been used to assess human proportionality or how one part of the body compares to another part. The interpretive and mathematical difficulties in this approach have been discussed in previous papers (Ross, Grand, Marshall, & Martin, 1984; Ross & Ward, 1982; Ross, Ward, Sigmon, & Leahy, 1983; Ross & Wilson, 1974). We have yet to find a single example where the traditional indices have any advantage whatsoever over the use of a single reference human as an imaging device. We contend that a ratio, being an inextricable combination of the variance in the numerator and denominator, is simply inappropriate in correlational analyses. Apart from historical reasons, the following general formula, with calculation options, replaces all the traditional indices and has the advantage

```
*****************************************************
    PHYSIQUE  MANAGEMENT  SYSTEMS  Inc.
         O-SCALE  PROGRAM
             Version 2.3
         Written by:  Richard Ward
           Updated March 1985


*****************************************************
```

```
        ----------------------------------------
        O-SCALE RATING FOR BODYBUILDER
          MAY 7TH
        ----------------------------------------

        MALE 25 YEARS OF AGE

        HEIGHT = 171.2     WEIGHT = 87.2

        PROPORTIONAL WEIGHT = 85.6 KG.

        SUM OF SKINFOLDS = 32.6 MM

        CORRECTED SUM OF SKINFOLDS = 32.4 MM

        I  I  I  I  I  I  I  I  I  I  I  I
        I.1.I.2.I.3.I.4.I.5.I.6.I.7.I.8.I.9.I

        A *

        W                                    *

        I...I...I...I...I...I...I...I...I...I
        4%.11%.23%.40%.60%.77%.89%.96%
```

PHYSIQUE ANALYSIS FOR BODYBUILDER
```
*****************************************************
```

ANTHROPOMETRIC MEASUREMENTS OF BODYBUILDER		SAME AGE/SEX NORM	
		4%	96%
WEIGHT (KG)	87.2	60.4	100
HEIGHT (CM)	171.2	167	190.9
TRICEPS SF (MM)	4.8	4.2	19.5
SUBSCAPULAR SF	7.2	7.1	27.8
SUPRASPINALE SF	3.8	4.2	27
ABDOMINAL SF	5.4	8.3	36.9
FRONT THIGH SF	6.6	6.6	24.1
MEDIAL CALF SF	4.8	3.7	18.4
ARM GIRTH (CM)	42.8	25.3	34.8
CALF GIRTH	39	32.7	42.1

```
        ----------------------------------------
```

PROPORTIONALITY PROFILE OF BODYBUILDER
VERSUS SAME AGE/SEX NORM

```
                4%              96%
        ----------------------------------------------
                 I               I
WEIGHT (KG)                      *
                 I               I
TRICEPS SF (MM)  *               I
                 I               I
SUBSCAPULAR SF   *               I
                 I               I
SUPRASPINALE SF  *               I
                 I               I
ABDOMINAL SF     *               I
                 I               I
FRONT THIGH SF *                 I
                 I               I
MEDIAL CALF SF   *               I
                 I               I
C.ARM GIRTH      I               I
                 I               I
C.CALF GIRTH     I               *
                 I               I
        ----------------------------------------------
```

Figure 6. O-Scale microcomputer printout for participant body builder in 1981 world championship.

of serving as an imaging device as we illustrate in the section on iconometrographics:

$$z = 1/s \ [v(170.18/h)^d - p]$$

where:

z   is a z-value or proportionality score

s   is a Phantom standard deviation for variable (v)

v   is any Phantom specified length, breadth, girth, skinfold thickness, mass, or volume

170.18   is the Phantom height constant (note: any variable may be optionally substituted as the scaling constant)

h   is the subject's obtained height (or optional scaling variable)

d   is the dimensional exponent (note: in the geometric similarity system all lengths, breadths, girths, and skinfold thicknesses, d = 1; all areas, surface and cross-sectional, d = 2; all weights and volumes of parts and the whole body, d = 3; in other dimensional systems or within sample analyses, the d-values can be assigned to accommodate various theoretical rationale as discussed by Ross et al. (1980)

p   is the given Phantom value ascribed for variable (v).

The Phantom as a Metaphorical Model

The Phantom is an arbitrary reference human. It is a calculation device, not a norm. The Phantom standard deviation(s) were derived from coefficients of variation from large male and female samples reported in the literature. By definition and central limit theory, the distribution of proportionality scores of the model is unimodal and symmetrical. The Phantom essentially size adjusts and scales sample data. It does not normalize or transform the data. It quantifies any departure from the Phantom p value in terms of standard scores. Because it is designed to be sensitive to small proportionality differences, it is also vulnerable to measurement error, especially systematic error among different investigators whose techniques have not been standardized. As for any standard score, the Phantom z-values may be treated arithmetically. Theoretical aspects and elucidations have recently been made in a paper incorporated in a testimonial publication in honor of Barbara Honeyman Heath-Roll (Ross, 1985).

Iconometrographics

It seems only yesterday, or perhaps it was the day before, it was axiomatic that in the design of experiments one must "define and delimit." It is small wonder that with limited calculation power, Quetelet and all those who followed in his tradition relied on simple ratios and the attractive two-compartment densitometric model with its simple prediction formula. However, with almost unlimited computer and graphic possibilities today, it is no longer a necessity to be parsimonious in data assembly, resolution, and report. The advent of the microcomputer emancipates the human biologist from drudgery in analysis and also frees him or her from the other alternative: dependency on the mainframe computer and subtle administrative and financial control. It also brings with it a need for technical excellence in measurement.

The new era should see increasing use of iconometrographic analyses of complex data assemblies on inexpensive microcomputers which are becoming as available as telephones. Identified by Boyd (1980) as a method of growth analysis, *iconometrographics* (as the term suggests) makes use of icons, or images, which are quantified and represented graphically. To illustrate iconometrographic analyses, to challenge anyone using traditional ratios to extract more meaning, and to report the persistence of proportional ethnic differences in the 1976 Montreal Olympic Games data, we compared 12 white and 27 black runners and boxers in profiles adjusted for differences in numbers in events, as shown in Figure 7.

By convention, when sample standard error bars do not overlap, the inference is that the difference between sample means is significant at about the .05 probability level. By this criterion, the blacks, represented by the unbroken line defining one standard error on either side of the mean, were shown to have proportionally longer forearms, shorter sitting height, and longer tibial height than the whites. They were also proportionally smaller in femur width and wrist girth. While we have shown blacks tending to have proportionally narrower hip widths than whites, we were not able to demonstrate this in these samples. The standard error bars for biiliocristal breadth overlapped, hence, we had to accept the null hypothesis. This was reflected in a nonsignificant $t = 1.24, p = .23$. Nevertheless, even in these limited samples where there is extreme selectivity in terms of physical performance, we were still able to show the persistence of ethnic specificity which we have demonstrated in larger samples using data from three Olympic Games (Ross & Ward, 1984).

Epilogue

The fate of any model in science is demise. Archimedes' principle for the prediction of percent fat and the Quetelet index simply account for less of the morphological variation in the human species than we need to know for scientific and practical purposes. To quantify adiposity and ponderosity related to appropriate norms for age and sex, we proposed the O-scale system. In pressing for the use of a more rational way of looking at human proportionality and fractionating body mass, we used simple sums of Phantom z-values to account for adipose tissue, bone, muscle, and residual tissue masses (Drinkwater & Ross, 1980). The fractionation of masses shown in Figure 7 depicts the proportional characteristics in a general way. However, our preference now is to use a five-way fractionation based on geometric derivation of regional masses from "shape constants" and the ascription of density based on cadaver evidence to yield estimated skin, adipose tissue, muscle, bone, and residual masses. These, when summed, account for total body mass in males and females over a wide age range (Drinkwater, 1984).

Much as walking may be thought of as a process of falling forward and reestablishing stability with a stride, so, too, do we advance in science. With

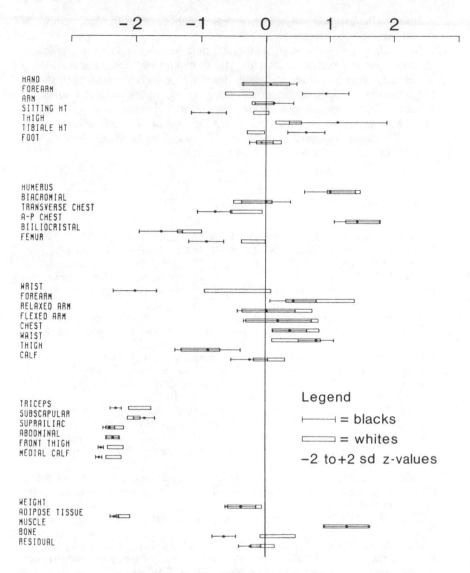

Figure 7. Proportionality profile of white and black runners and boxers. (▭ = white, ⊢——⊣ = black).

the most recent falling forward, we need now to reestablish stability by doing the following:

1. assessing compressibility effects in skinfold estimates of the adipose tissue and further defining the relationship between subcutaneous adipose tissue and internal adipose tissue masses using ultrasound, CAT scanning, and other medical imaging techniques;

2. fine-tuning estimates of bone mass using noninvasive, in vivo methods of measuring bone density;
3. expanding the quantitative anatomical data base by encouraging research using unembalmed cadavers with priority given to (a) obese males, (b) lean females, and (c) children and youth. Ideally all such basic anatomical studies should have coordinated medical imaging and histochemical studies.
4. testing our proposed new anthropometric models and other propositions in diverse in vivo samples, accounting for total body mass by independent derivation of fractional masses, and following longitudinal changes with growth, aging, disease, diet, and exercise.

References

Adams, J., Mottola, M., Bagnall, K.M., & McFadden, K.D. (1982). Total body fat content in a group of professional football players. *Canadian Journal of Applied Sport Sciences, 17*(1), 36-40.

Babu, D.S., & Chuttani, C.S. (1979). Anthropometric indices independent of age for nutritional assessment in school children. *Journal of Epidemiology and Community Health, 33*, 177-179.

Benn, R.T. (1971). Some mathematical properties of weight-for-height indices used as measures of adiposity. *British Journal of Preventive and Social Medicine, 1*(25), 42-50.

Billewicz, W.Z., Kemsley, W.F.F., & Thompson, A.M. (1962). Indices of obesity. *British Journal of Preventive and Social Medicine, 16*, 183-188.

Billewicz, W.Z., Thompson, A.M., & Fellowes, H.M. (1983). Weight for height in adolescence. *Annals of Human Biology, 10*(2), 119-124.

Boyd, E. (1980). In B.S. Savara & J.F. Schilke (Eds.), *Origins of the study of human growth*. Portland: University of Oregon Health Sciences Center Foundation.

Cerovska, J., Petrasek, R., Hajino, K., & Kaucka, J. (1977). Indices of obesity and body composition in four groups of Czech population. *Zeitschrift fur Morphologie Anthropologie, 68*(2), 213-219.

Colliver, J.A., Stuart, F., & Arthur, F. (1983). Similarity of obesity indices in clinical studies of obese adults: A factor analytic study. *American Journal of Clinical Nutrition, 38*(4), 640-647.

Conant, J.B. (1947). *On understanding science*. New Haven, CT: Yale University Press.

Cronk, C.E., Roche, A.F., Kent, R., Berkey, C., Reed, R.B., Valadian, I., Eichorn, D., & McCammon, R. (1982). Longitudinal trends and continuity in weight/stature^2 from 3 months to 18 years. *Human Biology, 54*(1), 729-749.

Drinkwater, D.T., & Ross, W.D. (1980). Anthropometric fractionation of body mass. In M. Ostyn, G. Beunen, & J. Simons (Eds.), *Kinanthropometry II* (pp. 177-189). Baltimore: University Park Press.

Drinkwater, D.T. (1984). *An anatomically derived method for the anthropometric estimation of human body composition*. Unpublished doctoral dissertation. Simon Fraser University, Burnaby, British Columbia.

Florey, C.D.V. (1970). The use and misuse of ponderal index and other weight-height ratios in epidemiological studies. *Journal of Chronic Diseases, 23*, 93-103.

Frisancho, A.R., & Flegel, P.N. (1982). Relative merits of old and new indices of body mass with reference to skinfold thickness. *American Journal of Clinical Nutrition, 36*, 697-699.

Garn, S.M., & Pesick, S.D. (1982). Comparison of the Benn Index and other body mass indices in nutritional assessments. *American Journal of Clinical Nutrition,* **36,** 573-575.

Goldbourt, U., & Medalie, J. (1974). Weight-height indices. *British Journal of Preventive and Social Medicine,* **16,** 116-126.

Hogben, L. (1957). *Statistical theory: The relationship of probability, credibility and error.* London: Allen and Unwin.

Jelliffe, D.B., & Jelliffe, E.P.F. (1979). Unappreciated pioneers, Quetelet: Man and index. *American Journal of Clinical Nutrition,* **32,** 2519-2521.

Katch, F.I. (1969). Practice curves and errors of measurement in estimating underwater weight by hydrostatic weighing. *Medicine and Science in Sports,* **1**(4), 212-220.

Keys, A., Fidanza, F., Karvonen, M.J., Kimura, N., & Taylor, H.L. (1972). Indices of relative weight and obesity. *Journal of Chronic Diseases,* **2**(25), 329-343.

Khosla, T., & Lowe, C.R. (1967). Indices of obesity derived from body weight and height. *British Journal of Preventive and Social Medicine,* **21,** 122-128.

Knowler, W.C., & Garrow, J. (1982). Obesity indices derived from weight and height. *International Journal of Obesity,* **6,** 241-243.

Lee, J., & Kolonel, L.N. (1983). Body mass indices: A further commentary. *American Journal of Clinical Nutrition,* **38,** 660-661.

Lee, J., Kolonel, L.N., & Hinds, M.W. (1981). Relative merits of the weight-corrected-for-height indices. *American Journal of Clinical Nutrition,* **34,** 2521-2529.

Lee, J., Kolonel, L.N., & Hinds, M.W. (1982). The use of an inappropriate weight-height derived index of obesity can produce misleading results. *International Journal of Obesity,* **6,** 233-239.

Malina, R.M., Meleski, B.W., & Shoup, R.F. (1982). Anthropometric, body composition and maturity characteristics of selected school-age athletes. *Pediatric Clinics of North America,* **29,** 1305-1323.

Martin, A.D. (1984). *An anatomical basis for assessing human body composition: Evidence from 25 cadavers.* Unpublished doctoral dissertation, Simon Fraser University, Burnaby, British Columbia.

Matiegka, J. (1921). The testing of physical efficiency. *American Journal of Physical Anthropology,* **4,** 223-230.

Quetelet, A. (1836). *Sur l'homme et le development des facultes.* Brussels: Hauman.

Roche, A.F., Siervogel, R.M., Chumlea, W.C., & Webb, P. (1981). Grading body fatness from limited anthropometric data. *American Journal of Clinical Nutrition,* **34,** 2831-2838.

Rolland-Cachera, M.F., Sempe, M., Guilloud-Bataille, M., Patois, E., Pcquignot-Guggenbuhl, G., & Fautrad, V. (1982). Adiposity indices in children. *American Journal of Clinical Nutrition,* **36,** 178-184.

Ross, W.D. (1985). Phantom stratagem for proportional growth assessment: Questions and answers. In O.G. Eiben (Ed.), *Human biologica budapestinensis.* A festival in honor of Barbara Honeyman Heath-Roll. **16,** 153-157.

Ross, W.D., Drinkwater, D.T., Bailey, D.A., Marshall, G.R., & Leahy, R.M. (1980). Kinanthropometry: Traditions and new perspectives. In M. Ostyn, G. Beunen, & J. Simons (Eds.), *Kinanthropometry II* (pp. 3-27). Baltimore: University Park Press.

Ross, W.D., Grand, T.I., Marshall, G.R., & Martin, A.D. (1984). On human and animal geometry. In M.L. Howell & D.N. Wilson (Eds.), *Proceeding of VII Commonwealth and International Conference on Sport, Physical Education, Recreation and Dance: Kinesiological sciences, Book 7* (pp. 77-97). Brisbane: Department of Human Movement Studies.

Ross, W.D., & Marfell-Jones, M.J. (1982). Kinanthropometry. In J.B. MacDougall, H.J. Green, & H.A. Wenger (Eds.), *Physiological testing of elite athletes.* Ottawa: Canadian Association of Sport Sciences.

Ross, W.D., & Ward, R. (1982). Human Proportionality and sexual dimorphism. In R.L. Hall (Ed.), *Sexual dimorphism in Homo Sapiens* (pp. 317-361). New York: Praeger.

Ross, W.D., & Ward, R. (1984). Proportionality. In J.E.L. Carter (Ed.), *The physical structure of Olympic athletes: Part II. Kinanthropometry of Olympic athletes: Vol. 19. Medicine and sport* (pp. 110-143). Basel: Karger.

Ross, W.D., & Ward, R. (1985). *The O-Scale system: Instructional guide for health and fitness professionals.* Surrey: Rosscraft.

Ross, W.D., Ward, R., Sigmon, B.A., & Leahy, R.M. (1983). Anthropometric concomitants of X-chromosome aueuploidies. In A.V. Sandberg (Ed.), *Cytogenetics of the mammalian X-chromosome, Part B: X-chromosome anomalies and their clinical manifestation* (pp. 127-157). New York: Alan R. Liss.

Ross, W.D., & Wilson, N.C. (1974). A stratagem for proportional growth assessment. In J. Borms & M. Hebbelinck (Eds.), *Children in Exercise. VI International Symposium on Pediatric Work Physiology.* ACTA Paediatrica Belgica (Suppl. 28), 169-182.

Siri, W.E. (1961). In J. Brozek & A. Henschel (Eds.), *Techniques for measuring body composition.* Washington, DC: National Academy of Science and National Research Council.

Tanner, J.M. (1976). Population differences in body size, shape and growth rate: A 1976 review. *Archives of Disease in Childhood, 51*(1), 1-2.

Visweswara, R.K., & Singh, D. (1970). An evaluation of the relationship between nutritional status and anthropometric measurements. *American Journal of Clinical Nutrition, 23*, 83.

Wilmore, J.H. (1983). Body composition in sport and exercise: Directions for future research. *Medicine and Science in Sports and Exercise, 15*(1), 21-31.

20

Validation by Cadaver Dissection of Matiegka's Equations for the Anthropometric Estimation of Anatomical Body Composition in Adult Humans

Donald T. Drinkwater, Alan D. Martin, and William D. Ross
SIMON FRASER UNIVERSITY
BURNABY, BRITISH COLUMBIA, CANADA

Jan P. Clarys
VRIJE UNIVERSITEIT BRUSSEL
BRUSSELS, BELGIUM

In 1921 the Czechoslovakian anthropologist Jindřich Matiegka proposed a comprehensive method for the anthropometric estimation of the weights of skin plus subcutaneous adipose tissue, skeletal muscle, bone, and "remainder" tissues (organs and viscera) in the human body. Although several investigators have acknowledged Matiegka's contribution (Behnke & Wilmore, 1974; Brožek, 1960; Drinkwater & Ross, 1980; Pařízková, 1977), this anthropometric approach has been largely neglected because of the popularity of methods such as densitometry for the estimation of body lipid and, until recently, the lack of suitable cadaver data to validate the method.

This work supported in part by Natural Sciences and Engineering Research Council of Canada Grant: A9402 and Fitness and Amateur Sports Council of Canada.

221

Matiegka developed a series of equations using groups of surface anthropometric measurements that closely relate to specific tissues. Using estimates of tissue weights inferred from cadaver data available to him at that time (Vierordt, 1906), Matiegka derived a series of coefficients for his equations which related these tissue weight estimates to surface measurements. He considered these coefficients to be only first approximations and acknowledged that additional cadaver evidence was required to validate them. The equations Matiegka developed were as follows:

(a) *Estimation of body weight*

$$W = D + M + O + R$$

where:

W   is body weight (g)
D   is skin + subcutaneous adipose tissue weight (g)
M   is muscle weight (g)
O   is skeletal weight (g)
R   is the remainder, or residual weight (g): organs, viscera, and all other tissues and fluids not accounted for by the other three components

(b) *Estimation of bone (skeletal) weight*

$$O = o^2 \times L \times k1$$

where:

O   is skeletal weight (g)
o   is $(o1 + o2 + o3 + o4)/4$, o1 to o4 being the maximal transverse diameters (cm) of the (o1) humeral and (o2) femoral condyles, (o3) wrist, and (o4) ankle breadths measured on one side of the body
L   is stature (cm)
k1 is 1.2

(c) *Estimation of skin-plus-subcutaneous adipose tissue weight*

$$D = d \times S \times k2$$

where:

D   is skin-plus-subcutaneous adipose tissue weight (g)
d   is $1/2 \times (d1 + d2 + d3 + d4 + d5 + d6)/6$, d1 to d6 being the thickness of skinfolds (mm) at the following sites: (d1) upper arm above the biceps; (d2) plantar side of the forearm at the level of maximum breadth; (d3) thigh above the quadriceps muscle, halfway between the inguinal fold and the knee; (d4) calf of the leg at the maximum girth; (d5) thorax on the costal margin halfway between the nipples and the umbilicus; and (d6) on the abdomen half-way between the navel and the anterior superior iliac spine.

S   is surface area (cm²), where

$$S = 71.84 \times W^{0.425} \times L^{0.725}$$

k2   is 0.13

(d) *Estimation of muscle weight*

$$M = r^2 \times L \times k3$$

where:

M   is (skeletal) muscle weight (g)
r   is (r1 + r2 + r3 + r4)/4, r1 to r4 being the average radii (cm) of the extremities without skin and subcutaneous adipose tissue as determined from circumferences and skinfolds measured on (r1) the flexed, unstrained arm above the belly of the biceps; (r2) the forearm at the maximum girth; (r3) the thigh half-way between the trochanter and the lateral epicondyle of the femur; (r4) the leg at the maximum circumference of the calf
L   is stature (cm)
k3   is 6.5

(e) *Estimation of organs plus viscera (remainder) weight*

$$R = W \times k4$$

where:

R   is the remainder weight (g) (i.e., the weight of organs, viscera, and all other tissues or fluids not otherwise accounted for)
W   is body weight (g)
k4   is 0.206

Method

Data to validate the method of Matiegka was derived from a cadaver dissection study undertaken at the Vrije Universiteit Brussel from October 1979 to June 1980. In all, 25 cadavers, aged 55 to 94 years, were measured, then dissected. This study was unique in that both extensive surface anthropometry and anatomical composition data were collected on the same cadavers.

This study (a) added to existing data on the quantities of tissues and organs in the adult human body, (b) related the quantities of these tissues and organs to external body measurements, and (c) yielded data which could be used both for validating various in vivo methods for the estimation of human body composition and for developing new anthropometric methods. A more complete description of the study may be found in Clarys, Martin, and Drinkwater (1984), and full details of methodology and listings of obtained data may be found in theses by Drinkwater (1984) and Martin (1984).

Results

Evaluation of Original Coefficients

Matiegka's equations were tested by applying them to cadaver data from the Brussels Study for which complete data existed (2 males, 3 females, embalmed; 6 males, 7 females, unembalmed). Only right-side surface measurements were used. The results are summarized in Table 1. Overall, skin-plus-subcutaneous adipose tissue and remaining tissue weights were substantially underestimated (-21.9% and -11.6%, respectively) whereas bone weight was substantially overestimated ($+24.8\%$). Muscle weight was somewhat underestimated (-8.5%), as was total body weight (-8.7%).

Determination of New Coefficients

To determine whether the predictive accuracy of the equations could be improved, new coefficients were calculated for each of the tissue groups using data from the 13 unembalmed cadavers. The coefficients determined were $k1$ = .17 (*SD* .05) for skin-plus-subcutaneous adipose tissue, $k2$ = 7.11 (*SD* .53) for muscle, $k3$ = .92 (*SD* .04) for bone, and $k4$ = 2.35 (*SD* .03) for the remainder, as compared to Matiegka's original coefficients: $k1$ = .13, $k2$ = 6.5, $k3$ = 1.2, and $k4$ = 2.03. The new coefficients were substituted into the equations and the equations reapplied to data of both embalmed and unembalmed cadavers. The results are summarized in Table 2.

Although the new coefficients generally resulted in an overall improvement of predictive accuracy (both sexes combined), the differences in error of prediction between males and females should be noted: (a) on average, skin-plus-subcutaneous adipose tissue of females was substantially overestimated ($+15.9\%$), whereas for males it was substantially underestimated (-12.3%); (b) muscle was slightly overestimated in females ($+2.7\%$) and slightly underestimated in males (-3.2%); and (c) conversely, the remainder was underestimated in females (-3.5%) and overestimated in males ($+5.4\%$). Bone was underestimated for both males (-4.9%) and females (-3.4%). Overall, total body weight was overestimated in females ($+4.5\%$) but underestimated in males (-3.3%). If only the results for the unembalmed cadavers are examined, the errors are lower. Gender-specific coefficients were also determined. Substitution of these coefficients into the equations resulted in a slight reduction of the mean percent error. However, because there was little improvement in the percent error variability across all tissues, these coefficients are not reported.

Discussion

Despite the limited data available to Matiegka, his coefficients are remarkably similar to those determined directly from the Brussels cadaver data. However, application of his original coefficients did lead to substantial errors in the prediction of the skin-plus-subcutaneous adipose tissue and bone. The new coefficients derived from the Brussels cadaver data could provide more ac-

Table 1. Differences (% error) between estimated and true tissue weights using Matiegka's equations with original coefficients for male and female cadavers

Sex	Group	Skin-plus-subcutaneous adipose			Skeletal muscle			Bone			Remainder			Total body weight		
		n	M	SD	n	M	SD	n	M	SD	n	M	SD	n	M	SD
M	Embalmed	2	−42.8	0.4	3	−13.1	7.4	6	19.0	7.5	6	−10.0	6.4	2	−14.9	4.8
	Unembalmed	6	−30.7	19.4	6	−10.7	5.6	6	28.5	6.4	6	−5.1	10.4	6	−9.6	4.2
	Both	8	−33.7	17.3	9	−11.5	5.9	12	23.7	8.3	12	−7.5	8.6	8	−10.9	4.7
F	Embalmed	3	−20.5	41.8	4	−6.5	14.6	6	18.3	12.9	6	−13.8	7.0	3	−10.0	6.0
	Unembalmed	7	−8.9	15.9	7	−5.8	7.8	7	32.1	6.7	7	−16.8	7.4	7	−5.7	4.4
	Both	10	−12.4	24.3	11	−6.1	10.0	13	25.7	12.0	13	−15.4	7.1	10	−7.0	5.0
All		18	−21.9	23.5	20	−8.5	8.7	25	24.8	10.2	25	−11.6	8.7	18	−8.7	5.1

Note. Data from Brussels Cadaver Analysis Study (1979-1980).

Table 2. Accuracy (% error) of tissue weight predictions using Matiegka's equations with revised coefficients for male and female cadavers

Sex	Group	Skin-plus-subcutaneous adipose			Skeletal muscle			Bone			Remainder			Total body weight		
		n	M	SD	n	M	SD	n	M	SD	n	M	SD	n	M	SD
M	Embalmed	2	-24.3	0.5	3	-5.0	8.1	6	-8.6	5.7	6	2.6	7.2	2	-8.4	4.8
	Unembalmed	6	-8.2	25.6	6	-2.3	6.2	6	-1.3	4.9	6	8.3	11.8	6	-1.6	6.0
	Both	8	-12.3	22.9	9	-3.2	6.5	12	-4.9	6.4	12	5.4	9.8	8	-3.3	6.2
F	Embalmed	3	5.2	55.4	4	2.2	16.0	6	-9.1	9.9	6	-1.7	8.0	3	0.8	10.7
	Unembalmed	7	20.5	21.1	7	3.0	8.5	7	1.5	5.2	7	-5.0	8.4	7	6.3	4.8
	Both	10	15.9	32.1	11	2.7	11.0	13	-3.4	9.2	13	-3.5	8.1	10	4.6	6.9
All		18	3.4	31.1	20	0.1	9.5	25	-4.1	7.8	25	0.8	9.9	18	1.1	7.6

Note. Values indicate differences (% error) between estimated and true tissue weights for male and female cadavers in the Brussels Cadaver Analysis Study (1979-1980).

curate predictions overall, but the coefficients so derived are to a large extent specific to that sample. It is probable that the true values of the coefficients are between those shown. Regardless of which set of coefficients is used, the error variability (*SD* percent error) about each tissue is rather large. A visual comparison of this variability for each tissue, whether predicted from the original coefficients or from the revised coefficients, reveals little difference. This is to be expected, because the new coefficients represent only the adjustment of a constant in the equations, which does not affect the variance. The original equations might be further improved if other measurement items are introduced.

Caution must be exercised when using these equations with the new coefficients to predict specific tissue weights for individuals rather than groups. These coefficients have been determined using a sample of old cadavers and may be inappropriate when applied to a living younger population, especially children. Nevertheless, the results do indicate that Matiegka's equations incorporating the new coefficients may be used to provide reasonable estimates of bone and muscle in the adult body, whereas estimates of skin-plus-subcutaneous adipose tissue and remaining tissues are far less certain.

References

Behnke, A.R., & Wilmore, J.H. (1974). *Evaluation and regulation of body build*. New Jersey: Prentice-Hall.

Brožek, J. (1960). The measurement of body composition. In M.F.A. Montagu, *A handbook of anthropometry* (pp. 78-136). Springfield, IL: Charles C. Thomas.

Clarys, J.P., Martin, A.D., & Drinkwater, D.T. (1984). Gross tissue weights in the human body by cadaver dissection. *Human Biology, 56*(3), 459-473.

Drinkwater, D.T., & Ross, W.D. (1980). The anthropometric fractionation of body mass. In M. Ostyn, G. Beunen, & J. Simons (Eds.), *Kinanthropometry II* (pp. 178-189). Baltimore: University Park Press.

Drinkwater, D.T. (1984). *An anatomically derived method for the anthropometric estimation of human body composition*. Unpublished doctoral dissertation, Simon Fraser University, Burnaby, British Columbia.

Martin, A.D. (1984). *An anatomical basis for assessing human body composition: Evidence from 25 dissections*. Unpublished doctoral dissertation, Simon Fraser University, Burnaby, British Columbia.

Matiegka, J. (1921). The testing of physical efficiency. *American Journal of Physical Anthropology, 4*(3), 223-230.

Pàrízková, J. (1977). *Body fat and physical fitness*. The Hague: Nijhoff.

Vierordt, H. (1906). *Anatomische, physiologische und physikalische Daten und Tablellen*. Jena: G. Fisher.

21

Proportionality of Muscle Volume Calculations of Male and Female Participants in Different Sports

Tjaart S.P. Van der Walt
POTCHEFSTROOM UNIVERSITY FOR C.H.E.
POTCHEFSTROOM, REPUBLIC OF SOUTH AFRICA

Johannes H. Blaauw
UNIVERSITY OF STELLENBOSCH
STELLENBOSCH, REPUBLIC OF SOUTH AFRICA

Marius Desipré
UNIVERSITY OF THE ORANGE FREE STATE
BLOEMFONTEIN, REPUBLIC OF SOUTH AFRICA

Hans O. Daehne
UNIVERSITY OF PRETORIA
PRETORIA, REPUBLIC OF SOUTH AFRICA

Johan P. van Rensburg
CHAMBER OF MINES
JOHANNESBURG, REPUBLIC OF SOUTH AFRICA

In any analysis of muscle volume measurements using anthropometric techniques, certain error can be expected. For example, the assumption made by Wartenweiler, Hess, and Wüest (1974) where limb segments are treated as cylindrical obviously leads to volume measurements much larger than those expected. With a formula which takes the amount of tapering in limb segments into account (Van der Walt & Turnbull, in press), a clearer picture can emerge.

This study was made possible by a grant from the Department of National Education: Branch Sport Advancement of the Republic of South Africa and the South African Association for Sport, Physical Education and Recreation.

Absolute measurements, however, do not give a true picture when comparing participants in different sporting events. A solution to the problem can be found in proportionality assessment. In an excellent overview of the relevant facts on proportionality, Ross, Ward, Leahy, and Day (1982) stated that the goal with proportionality assessment is to obtain an appreciation of relative size. In this study, the Phantom stratagem as proposed by Ross and Wilson (1974) is used to compare limb segment muscle volumes of athletes in different sporting events.

Method

A total of 830 male and female participants in the South African Festival Games of 1981 were measured during the South African Games Anthropometric Project (SAGAP). In this report, data on 562 male participants in 30 sporting events and on 158 female participants in 13 events are reported. All measurements were taken according to the methods applied during the Montreal Olympic Games Anthropological Project (MOGAP) (Carter, 1982). Anthropometric instruments were standard and comparable to those used in MOGAP. Measurements of body mass, stature, acromial height, radial height, stylion height, tibial height, sitting height, biepicondylar breadth, bicondylar breadth, upper arm girth (extended and relaxed), forearm girth, wrist girth (distal), thigh girth, calf girth, ankle girth (proximal), and skinfolds of triceps, biceps, forearm, thigh (front), and calf (medial) were made. Lengths were derived from projected heights as reported by Carter (1982).

All volumetric calculations are given in cubic centimeters. Calculations of the volumes of the upper arm were done according to the method reported by Wartenweiler et al. (1974), assuming the limb segment to be cylindrical (Figure 1a). Bone volumes and skinfold thicknesses were subtracted to obtain lean muscle volumes. Bone volumes were calculated using a constant of 3.0 designating the relation of the epiphyseal diameter (epicondylar) to the diaphysial (shaft) diameter. Calculations of the muscle volumes of the forearm, thigh, and calf assumed a geometric shape described as a frustrum of a right circular cone with two radii (Figure 1b).

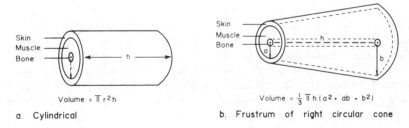

Volume $= \pi r^2 h$

a. Cylindrical

Volume $= \frac{1}{3} \pi h (a^2 + ab + b^2)$

b. Frustrum of right circular cone

Figure 1. Geometrical shapes used for calculations of muscle volumes.

To calculate radius a in Figure 1b, the wrist girth, bicondylar diameter, and ankle girth were used for forearm, thigh, and calf muscle volume calculations, respectively. Radius b was calculated using forearm girth, thigh girth, and calf girth measurements. Bone volumes were calculated as described above, setting bone diameters for the forearm equal to those of the upper arm and for the calf equal to that of the thigh.

For calculations of the z values, the formula reported by Ross et al. (1982) was used. Because no Phantom values for the muscle volumes reported here were available, the mean values and standard deviations of the total SAGAP sample (Table 1) were used thusly:

$$z = \frac{1}{s} [v(\frac{174.14}{h})^3 - p]$$

where s is the standard deviation of the SAGAP sample for a variable, 174.14 is the mean stature for the SAGAP sample, v is the muscle volume, h is the subject's stature, and p is the mean SAGAP value for that muscle volume.

Comparisons between different sporting events were made using proportionality profiles with bar graphs set at two standard errors of the mean. The use of such profiles is explained by Ross et al. (1982). Differences between means of muscle volume measurements were also calculated after the measurements had been adjusted for the Phantom stature as reported by Ross et al. (1982). A Bonferroni Test for pairwise comparisons was used, $\alpha_{EW} = .05$.

Table 1. Means, standard deviations, ranges, and standard errors of muscle volume measures for SAGAP participants

Measure	M	SD	Range	SE
Age (yr)	25.3	8.27	9.1- 71.9	0.29
Mass (kg)	70.4	13.15	26.2- 141.5	0.46
Stature (cm)	174.1	9.78	132.3- 220.0	0.34
Muscle vol. upper arm*	1884.0	599.38	455.6- 4095.1	20.80
Muscle vol. forearm*	1161.7	312.86	333.8- 2142.2	10.86
Muscle vol. upper extremity*	2995.1	852.92	807.4- 5613.0	29.61
Muscle vol. thigh*	9398.3	1894.70	3191.6-15660.6	65.77
Muscle vol. calf*	3926.0	884.12	1279.7- 7371.0	30.86
Muscle vol. lower extremity*	12820.6	2529.35	4303.4-20659.2	87.79
Tapered vol. forearm**	891.6	210.84	295.7- 1524.5	7.32
Tapered vol. upper extremity**	2670.1	768.43	736.5- 5302.7	26.67
Tapered vol. thigh**	6208.3	1116.13	2416.5- 9896.7	38.74
Tapered vol. calf**	2991.7	586.5	1123.0- 4919.9	20.37
Tapered vol. lower extremity**	8462.2	1510.5	3181.6-13239.1	52.46

Note. N = 830; *assumed shape right circular cylinder (cm³); **assumed shape frustrum of right circular cone (cm³).

For simplicity, the different sporting events were classified into four groups, that is, stunt-type events (3 female, 4 male events), water-activity events involving swimming (4 female, 9 male events), sports emphasizing leg activity (7 female, 10 male events), and sports mainly involving upper body activity (9 male events). Some sports, for example, gymnastics and ice skating, fall into more than one category.

Results

Table 1 represents the descriptive statistics for the total group measured in the SAGAP Study. Mean limb segment muscle volumes calculated either with a formula for a right circular cylinder or a frustrum of a right circular cone are also included in the table. The proportionality profiles of z values for limb segment volumes for males and females are represented in Figures 2 and 3. Differences in muscle volume measurements among athletes in different sports can be noted as well as the differences, or lack of them, in the muscle volume measurements among athletes in the same sport. An overlap between measurements (± 2 SE for the z value are shown) indicates no significant differences, whereas no overlap would show a tendency for significant differences. In Tables 2 and 3 significant differences among the Phantom adjusted mean muscle volumes for the different sporting groups are reported.

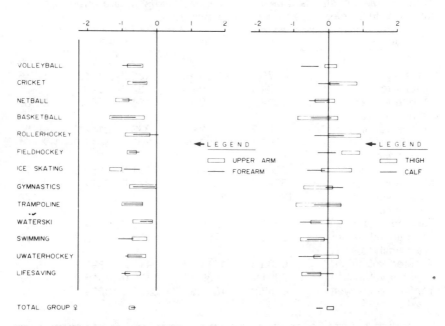

Figure 2. Proportionality profile of z values for limb segment muscle volumes in females.

Table 2. Differences among Phantom-adjusted muscle volume means (170.18/stature) for women's sport events[a]

Stunt-type activities

Body mass:	gym	tramp	ice-sk
Vol. upper arm:	ice-sk	tramp	gym
Vol. forearm:	tramp	ice-sk	gym
Vol. thigh:	tramp	gym	ice-sk
Vol. calf:	ice-sk	tramp	gym

Water activities

Body mass:	wtrski	swim	uwhock	lfesave
Vol. upper arm:	wtrski	uwhock	lfesave	swim
Vol. forearm:	swim	lfesave	uwhock	wtrski
Vol. thigh:	wtrski	swim	lfesave	uwhock
Vol. calf:	wtrski	uwhock	swim	lfesave

Leg activities

Body mass:	ice-sk	fhock	rolhock	baskbll	netball	cricket	volball
Vol. upper arm:	ice-sk	fhock	baskbll	netball	rolhock	cricket	volball
Vol. forearm:	ice-sk	baskbll	fhock	volball	cricket	netball	rolhock
Vol. thigh:	ice-sk	baskbll	volball	rolhock	cricket	fhock	netball
Vol. calf:	ice-sk	volball	fhock	rolhock	cricket	baskbll	netball

Note. Means are ordered low to high, left to right. Means not underlined by a common line differ significantly ($p < .05$) from each other; [a]see Figure 2 for sport events list.

Table 3. Differences among Phantom-adjusted muscle volume means (170.18/stature) for men's sport events[a]

Stunt-type activities

Body mass:	tramp	gym	ice-sk	dive
Vol. upper arm:	tramp	ice-sk	dive	gym
Vol. forearm:	tramp	ice-sk	dive	gym
Vol. thigh:	tramp	gym	ice-sk	dive
Vol. calf:	tramp	ice-sk	gym	dive

Leg activities

Body mass:	distrun	ice-sk	tennis	sprint	rolhock	cycling	icehock	fhock	baskbll	volball
Vol. upper arm:	distrun	ice-sk	fhock	tennis	sprint	cycling	icehock	baskbll	rolhock	volball
Vol. forearm:	distrun	ice-sk	icehock	sprint	fhock	cycling	tennis	baskbll	rolhock	volball
Vol. thigh:	distrun	ice-sk	baskbll	tennis	sprint	rolhock	icehock	fhock	cycling	volball
Vol. calf:	ice-sk	distrun	icehock	cycling	tennis	rolhock	fhock	volball	baskbll	sprint

(Cont.)

Table 3. (Cont.)

Swimming activities

Body mass:	surfing	wtrski	swim	sprfish	lfesave	scuba	bchlife	uwhock	wpolo
Vol. upper arm:	surfing	scuba	wtrski	sprfish	swim	bchlife	uwhock	lfesave	wpolo
Vol. forearm:	surfing	swim	scuba	sprfish	lfesave	wpolo	wtrski	uwhock	bchsave
Vol. thigh:	surfing	wtrski	swim	lfesave	scuba	bchlife	wpolo	sprfish	uwhock
Vol. calf:	surfing	wtrski	sprfish	scuba	uwhock	wpolo	swim	bchlife	lfesave

Upper body activities

Body mass:	boxing	gym	canoe	judo	rowing	wrestl	karate	bbuild	weightl
Vol. upper arm:	boxing	rowing	canoe	judo	karate	wrestl	gym	weightl	bbuild
Vol. forearm:	boxing	rowing	judo	wrestl	karate	canoe	weightl	gym	bbuild
Vol. thigh:	gym	boxing	judo	canoe	wrestl	rowing	karate	bbuild	weightl
Vol. calf:	judo	gym	boxing	canoe	rowing	wrestl	weightl	karate	bbuild

Note. Means are ordered low to high, left to right. Means not underlined by a common line differ significantly ($p < .05$) from each other; [a]see Figure 3 for sport events list.

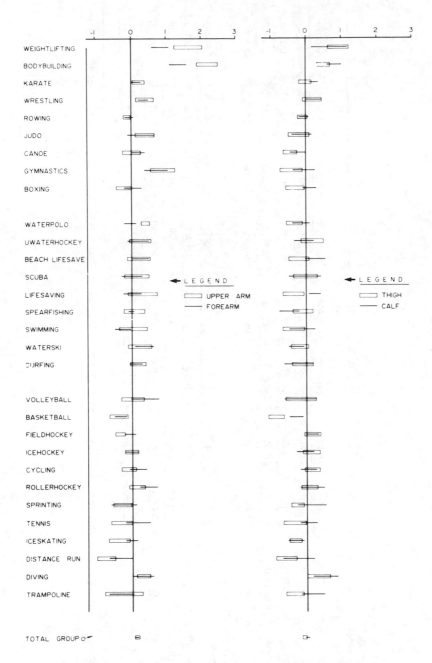

Figure 3. Proportionality profile of *z* values for limb segment muscle volumes in males.

Discussion

Women participants in the swimming events reported here do not differ significantly in muscle volume measurements (see Table 2), the exception being the difference in calf muscle volumes between the lifesavers and waterskiers. Exactly the same results for the same water events were found for men (Table 3), the only significant differences being calf volume measurements for lifesavers and waterskiers. The conclusion was that muscle volumes of the extremities tend to be the same for water events and that smaller muscle volumes are found among waterskiers (men and women) and surfers (men), with the only exception to this being the forearm muscle volumes of waterskiers.

In the stunt-type events the muscle volumes of the upper extremities in both men and women tend to be the largest for the gymnasts, with significant differences between men trampolinists and gymnasts (Table 3). Muscle volumes of the lower extremities show proportionally larger calf than thigh volumes for gymnasts and vice versa for ice skaters (Figures 2 & 3 and Tables 2 & 3). Divers (men) tend to have the larger muscle volumes in the lower extremities.

Among the leg activities, it is clear that the ice skaters (men and women) and the middle- and long-distance runners (men) have the lowest muscle volumes in both the upper and lower extremities (Tables 2 and 3). Larger thigh muscle volumes occurred among the men cyclists, roller hockey, field hockey, and volleyball players (Table 3, Figure 3), and among women hockey players and cricketers.

For the upper body activities (men) the weightlifters and body builders fall into their own category (Table 3, Figure 3). It should be noted that gymnasts, although rating second lowest in body mass, rank very high on upper extremity muscle volumes. Boxers rank surprisingly low on upper extremity muscle volumes. In general, the forearm muscle volume seems to be the one least affected by different activities, showing the fewest significant differences among sports. The proportionally largest differences in muscle development between the upper arm and forearm are those of wrestlers, bodybuilders, water polo players (men, Figure 3), and swimmers (women, Figure 2). The proportionally largest differences in muscle volume between calf and thigh are found among men lifesavers and basketball players (Figure 3), and among women volleyball players and field hockey players (Figure 2). Large but not significant differences between the same measurements were found in body builders, karatika, boxers, and trampolinists (men), and ice skaters (women).

The sporting events favoring larger upper limb and smaller lower limb development among men are judo, gymnastics, waterpolo, beach lifesaving, lifesaving, and waterskiing, and among women, gymnastics, swimming, and waterskiing. Sporting events favoring larger lower limb development and smaller upper limb development for men and women are field hockey and, to a lesser extent, roller hockey (women). Even when grouped with comparable events and when measurements are scaled up/down for stature, some sporting events still are associated with characteristic upper and lower extremity muscle volumes in competitors.

References

Carter, J.E.L. (Ed.). (1982). *Physical structure of Olympic athletes: Part 1. The Montreal Olympic Games Anthropological Project.* Basel: S. Karger.

Ross, W.D., & Wilson, N.C. (1974). A stratagem for proportional growth assessment. *Acta Paediatrica Belgica, 74* (Suppl. 28), 169-182.

Ross, W.D., Ward, R., Leahy, R.M., & Day, J.A.P. (1982). Proportionality of Montreal Athletes. In J.E.L. Carter (Ed.), *Physical structure of Olympic athletes: Part 1. The Montreal Olympic Games Anthropological Project* (pp. 81-106). Basel: Karger.

Van der Walt, T.S.P., & Turnbull, R. (in press). Lower extremity muscle volumes of South African middle and long distance male athletes. *South African Journal for Research in Sport, Physical Education and Recreation.*

Wartenweiler, J., Hess, A., & Wüest, B. (1974). Anthropologic measurements and performance. In L.A. Larson (Ed.), *Fitness, health and work capacity: International standards for assessment* (pp. 211-240). New York: McMillan.

22

Body Composition Estimation by Girths and Skeletal Dimensions in Male and Female Athletes

Wayne E. Sinning and Anthony C. Hackney
KENT STATE UNIVERSITY
KENT, OHIO, USA

Concern about the weight control of athletes has led to the popular use of skin-folds (SF) and other anthropometric measures in regression equations to estimate body composition components. It is known that the equations tend to be population specific; that is, they may show marked error when applied to a population which varies from the original sample in body composition or in anthropometric measures used. Athletes do form specialized subsamples within the population. In previous work, the accuracy of the new generalized curvilinear models of Durnin and Womersley (1974), Jackson and Pollock (1978), and Jackson, Pollock, and Ward (1980) was compared to linear models on samples of male athletes by Sinning et al. (1983) and by Sinning and Wilson (1984) on female athletes, while Thorland, Johnson, Tharp, Fagot, and Hammer (1984) reported similar research on adolescent athletes. Some of the generalized equations were very accurate, especially selected equations by Jackson and Pollock (1978) on men and Jackson, Pollock, and Ward (1980) on women.

The equations previously evaluated were based on SFs. Equations based on skeletal diameters (SK) and girths (G) are of interest because the equipment investment is much less than for high quality SF calipers. This investigation evaluated selected equations that use SK and G measures to estimate the body composition of male and female athletes.

Method

Subjects were 265 male college athletes representing 12 sports in a Division III NCAA college and 79 female athletes representing 9 sports in a Division I university. Measurements were taken during the competitive seasons except for men's soccer and football when they were taken during spring practice. Age (mean $\pm$ *SD*) of the female subjects was 19.8 $\pm$ 1.30 years while males were 20.2 $\pm$ 1.84 years. Weight, density (BD), and relative fat (% fat) values for females were 60.2 $\pm$ 7.51 kg, 1.052 $\pm$ 0.013 g/cm³, and 20.1 $\pm$ 5.3%. For males, respective values were 75.7 $\pm$ 10.61 kg, 1.079 $\pm$ 0.011 g/cm³, and 9.2 $\pm$ 4.4%. Body composition was computed from BD measured by underwater weighing (UWW) (Sinning, 1977). Residual volume (RV) was measured out of water for males by the oxygen dilution technique (Wilmore, 1969) and for females by helium dilution. The % fat was computed by the equation of Brozek, Grande, Anderson, and Keys (1963).

The measurements used in the regression equations are shown in Table 1. The sites used were those described by Behnke and Wilmore (1974, pp. 38-52). Three equations were evaluated, two for females and one for males (Table 1). Equations used for females were by Wilmore and Behnke (WBF) (1970), using wrist diameter with maximum abdominal, hip, arm, and forearm girths; and by Katch and McArdle (KMF) (1973), using the average abdominal, forearm, and thigh girths. For males, the Wilmore-Behnke equation (WBM) (1969) using weight and waist girth was employed. Estimates were converted

Table 1. Body dimensions and regression equations used to determine body composition

Measure	Ref. No.[a]	*M*	*SD*	*SE*	Range
Females (*N* = 79)					
Wrist diameter (cm)	1	5.1	0.26	0.03	4.6- 5.8
Arm girth extended (cm)	2	25.4	1.91	0.22	21.4- 31.6
Forearm girth (cm)	3	24.2	1.37	0.15	20.5- 28.8
Abdomen girth minimum (cm)	4	70.0	4.59	0.52	61.8- 86.4
Abdomen girth maximum (cm)	5	76.5	5.25	0.59	65.6- 95.7
Hip girth (cm)	6	93.2	5.64	0.63	82.8-117.3
Thigh girth (cm)	7	57.3	3.96	0.45	49.6- 73.0
Males (*N* = 265)					
Waist girth (cm)	8	81.4	7.12	0.44	67.3-111.0
Weight (kg)	9	75.7	10.61	0.65	55.6-119.3

Equations
Wilmore-Behnke females (WBF)
$$BD = 1.065351 + .0112X_1 - .00055X_5 - .00082X_6 - .00159X_2 + .00362X_3$$
Katch-McArdle females (KMF)
$$BD = 1.14465 - .0015(X_4 + X_5/2) + .00448X_3 - .00165X_7$$
Wilmore-Behnke males (WBM)
$$LBW = 44.636 + 1.0817X_9 - .7396X_8$$

Note. BD = body density, LBW = lean body weight; [a]the reference number is used to refer to the specific measurements throughout the text; for example, wrist diameter is X_1.

to % fat for comparison to the criterion measures. The equation of Brozek et al. (1963) was used when BD was estimated. When lean body weight (LBW) was estimated, the % fat was computed as follows: % fat = [1.00 − (LBW/weight)] × 100.

Estimated % fat values were compared to UWW measured values by the Student t-test. Regression analysis was used to evaluate relationships between true and estimated values. Statistics were interpreted at the .05 level of significance. The true error, which is the square root of the mean of the sum of squares of the residuals, was also computed ($E = \sqrt{\Sigma(Y - \hat{Y})^2/N}$, where Y is the criterion value and $\hat{Y}$ is the estimate). The E and the standard error of estimate (*SEE*) should be about the same value for acceptable equations (Lohman, 1981).

Results

The statistical analyses are summarized in Table 2. The estimated % fat by WBF was not significantly different from the UWW measured value. KMF significantly overestimated % fat by 4.3%. The mean WBM estimate was also significantly high. The regression analyses are given in Figure 1. All correla-

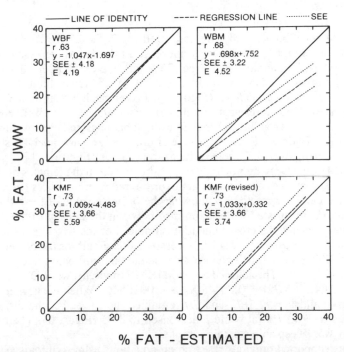

Figure 1. Regression analysis for the equations studied. WBF, Wilmore-Behnke females; WBM, Wilmore-Behnke males; KMF, Katch-McArdle females; KMF (Revised) revised KMF equation. Regression lines and standard error of estimate (*SEE*) are plotted only over the range of scores for the samples.

Table 2. Comparison between underwater weighing % fat values and values estimated from equations

Equation	N	M % fat	SD	Difference[a]	t
WBF	79	20.8	3.22	+0.7	1.56
KMF	79	24.4	3.89	+4.3	10.44*
WBM	265	12.1	4.30	+2.9	13.56*

Note. WBF = Wilmore-Behnke females; KMF = Katch-McArdle females; WBM = Wilmore-Behnke males; [a]compared to underwater weighing values, for females 20.1% fat, for males 9.2% fat; *p < .05.

tion coefficients were significant. For WBF, E was almost the same as the SEE (4.19 and 4.18% fat), but for KMF E was much larger (5.59 and 3.66% fat). For WBM, SEE was 3.22, while E was 4.52% fat. Slopes for the WBF and KMF equations approached unity (1.047 y/x and 1.009 y/x, respectively), while the slope for WBM was rotated toward the x axis (.698 y/x).

Discussion

The WBM equation markedly overestimated the % fat (Table 2). Regression analysis showed the overestimate to increase as the % fat increased. At the upper limit of the sample distribution (26% fat), the expected mean estimate from the regression equation would be 38.7%. At the lower limit (0.6% fat) the estimate would be 1.5% fat (Figure 1). E was much higher than SEE, reflecting the extreme overestimate of fat in the fatter subjects. The WBF equation met criteria for acceptability. There was a slight but nonsignificant overestimate of 0.7% fat (Table 2). E and SEE were similar. Regression analysis showed a uniform error over the range of measured values (Figure 1).

The KMF equations significantly overestimated fat in the females by 4.3%. The overestimate was almost constant over the range of measures (Figure 1). Such error can be corrected by adjusting the ordinate value of the estimation equation to compensate for the difference between the density of the derivation population and the cross-validation sample. For the Katch and McArdle (1973) sample, the density was 1.039 versus 1.052 g/cm³ for the present sample. The difference of 0.013 g/cm³ was added to their ordinate value changing it to 1.15725 g/cm³. The adjusted equation [KMF (revised)] was BD = 1.15725 − 0.0015 $(X_4 + X_5/2)$ + 0.00448 X_3 − 0.00165 X_7. When the new equation was cross-validated, r, SEE, and SD remained the same, but the difference decreased to −0.97% fat while E decreased to 3.75 (Figure 1). The revised equation was acceptable for this sample.

Results supported previous research on equations using skinfolds (Sinning & Wilson, 1984; Sinning et al., 1985; Thorland et al., 1984), showing that regression equations can provide acceptable estimates for body composition screening provided they are appropriately cross-validated. Durnin and Womersley (1974), Jackson and Pollock (1978), Jackson et al. (1980), and

Lohman (1981) noted the problem of curvilinearity between body fat and skinfold measures as well as the effect of age when using skinfolds to estimate BD and % fat. Similar evaluations of G and SK dimensions are not available. However, for skinfolds, these effects are of concern primarily when there is a wide age range in the sample and high levels of fatness. The ranges of both variables in the present sample were limited, and consequently, of less concern.

As in other studies (Sinning & Wilson, 1984; Sinning et al., 1985; Thorland et al., 1984), the present findings illustrate the need to cross-validate equations on specific populations before using them. Also, the error of the estimate may be due to the technique of measurement as well as to the underlying statistical assumptions. Such error and its implications have been discussed (Sinning & Wilson, 1984). In using the WBF and KMF equations, it must be remembered that the cross-validation is specific and applies only to athlete populations similar to the present samples in age, overall size, and fatness.

References

Behnke, A.R., & Wilmore, J.H. (1974). *Evaluation and regulation of body build and composition*. Englewood Cliffs, NJ: Prentice-Hall.

Brozek, J., Grande, F., Anderson, J.T., & Keys, A. (1963). Densitometric analysis of body composition: Revision of some quantitative assumptions. *Annals of the New York Academy of Science, 110*, 113-140.

Durnin, J.V.G.A., & Womersley, J. (1974). Body fat assessed from total body density and its estimation from skinfold thickness: Measurements on 481 men and women aged from 16 to 72 years. *British Journal of Nutrition, 32*, 77-97.

Jackson, A.S., Pollock, M.L., & Ward, A. (1980). Generalized equations for predicting body density of women. *Medicine and Science in Sports and Exercise, 12*, 175-182.

Jackson, A.S., & Pollock, M.L. (1978). Generalized equations for predicting body density of men. *British Journal of Nutrition, 40*, 497-504.

Katch, F.I., & McArdle, W.D. (1973). Prediction of body density from simple anthropometric measurements in college-age men and women. *Human Biology, 45*, 445-454.

Lohman, T.G. (1981). Skinfolds and body density and their relation to body fatness: A review. *Human Biology, 53*, 181-225.

Sinning, W.E. (1977). Body composition by body densitometry. In M. Adrian & J. Brames (Eds.), *NAGWS Research Reports: Vol. III* (pp. 138-152). Washington, DC: American Alliance for Health, Physical Education and Recreation.

Sinning, W.E., Dolny, D., Little, K.D., Cunningham, L., Racaniello, A., Siconolfi, S., & Sholes, J. (1985). Validity of "generalized" equations for body composition analysis in male athletes. *Medicine and Science in Sports and Exercise, 17*, 124-130.

Sinning, W.E., & Wilson, J.R. (1984). Validity of "generalized" equations for body composition analysis in women athletes. *Research Quarterly for Exercise and Sport, 55*, 153-160.

Thorland, W.G., Johnson, G.O., Tharp, G.D., Fagot, T.G., & Hammer, R.W. (1984). Validation of anthropometric equations for the estimation of body density in adolescent athletes. *Medicine and Science in Sports and Exercise, 16*, 77-81.

Wilmore, J.H. (1969). A simplified method for the determination of residual lung volumes. *Journal of Applied Physiology, 27*, 96-100.

Wilmore, J.H., & Behnke, A.R. (1970). An anthropometric estimation of body density and lean body weight in young women. *American Journal of Clinical Nutrition, 23*, 267-274.

Wilmore, J.H., & Behnke, A.R. (1969). An anthropometric estimation of body density and lean body weight in young men. *Journal of Applied Physiology, 27*, 25-31.

23

Hydrostatic Weighing of Women Throughout the Menstrual Cycle

Milan D. Svoboda and Lovina M. Query
PORTLAND STATE UNIVERSITY
PORTLAND, OREGON, USA

In assessing daily weight changes, a gain of one or two pounds is often interpreted by the lay person to mean an increase in fat. Yet changes in body weight (BW) can arise from other causes than an actual change in fat. For example, the weight changes reported by women during the course of the menstrual cycle have been documented to be due to water retention (Bruce & Russell, 1962; Good, 1978; Thorn, Nelson & Thorn, 1938) and/or appetite (Dalvit, 1981).

When measuring body composition by the method of hydrostatic weighing (HW), the hydration level of the subject is not monitored. Recent evidence has shown that variations in hydration level result in significantly different estimates of percent fat (% fat) and the derived estimates of fat weight (FW) and fat free weight (FFW) (Girandola, Wiswell, & Romero, 1977). The apparent changes in % fat due to variations in hydration are not real changes in body composition. They occur because of the assumptions made when using either of the commonly used formulae for predicting % fat from density (Brozek, Keyes & Anderson, 1963; Siri, 1961). These assumptions include: the density of fat is .9 kg/l and the density of the fat free mass is 1.1 kg/l. In other words, it is assumed that the only reason that overall body density (D) varies from individual to individual is because of differences in relative proportions of fat mass and fat free mass and not because of differences in the density of either component. Yet when the hydration level of the individual varies, the density of the fat free mass is influenced because water is one of its major components (Lohman, 1981). Lohman estimates that variability in water content of the body makes the largest contribution to the error in predicting % fat from D. The standard error is estimated by Lohman to be $\pm$

3.9% fat when applying HW techniques to any population other than adult males.

Whether the weight changes experienced by women throughout the course of the menstrual cycle will also result in differences in estimated % fat has not received much attention. One recent study found no significant differences in % fat by HW measured at three points during each of two menstrual cycles (Byrd & Thomas, 1983). However, with so few measurements, it is possible that HW was not performed at times of peak weight gain or loss. The present study was designed to eliminate this difficulty and to determine whether HW results in significantly different estimates of % fat in women at different times in the menstrual cycle.

Method

Twenty-six healthy women served as subjects. Volunteers were accepted as subjects only if they were not using intrauterine devices or oral contraceptives and were willing to do less than 1 and 1/2 hours of exercise per week. They agreed to avoid salted foods and/or simple sugars prior to each measurement. As soon as possible after Day 1 of their menstrual cycle (first day of flow), initial measures of BW and HW were performed. Thereafter, BW was measured three times per week at the same time of the day throughout one cycle. Hydrostatic weighings were done under similar conditions on every second BW trial. Additional HW measures were also made if (a) a subject gained 0.5 kg or more compared to the previous weighing, or (b) if the BW was higher (or lower) than all previous weighings. These additional measurements were taken to increase the probability of obtaining HW trials on the days when BW changes were extreme. One additional BW and HW trial was performed at the beginning of the next menstrual cycle, on the same day in the cycle as the initial measures were obtained. These overlap measures were to observe whether any net changes had occurred during the course of the cycle.

Body weight was measured on a Homs Balance Beam Scale accurate to $\pm$ 50 grams. Hydrostatic weighings were made in a 500-gallon tank using a Chatillon 9-kilogram scale. The first time that HW was performed, 10 trials were used; on the second day of HW, a minimum of 7 trials was performed; thereafter, trials were terminated when three similar weights or 10 trials were achieved. On all days the mean of the three highest weights within $\pm$ 50 grams was used as the subject's score for that day. Density (D) was calculated using the formula of Novak (1974), and Siri's formula (1961) was used to convert D to % fat.

Residual volume (RV) was measured out of the water using the revised oxygen dilution method (Wilmore, Vodak, Parr, Girandola, & Billing, 1980). Residual volume under ambient conditions was converted to body temperature, pressure, saturated (BTPS) units using standard procedures. Two RV trials were taken on a given day. For all but 2 subjects, RV was remeasured on one or two additional occasions during the subject's cycle. To eliminate the influence of variability in residual volume on estimates of % fat, the average of all measures of RV taken on a subject was used in all calculations of body density.

To allow comparisons to be made among women with menstrual cycles of differing length, the luteal phase was estimated as beginning 15 days before the onset of menstruation (Guyton, 1981). Measurements of BW and HW taken during this time period were broken into three blocks of 5 days each (Blocks 1 to 3, respectively, counting backwards from the onset of menstrual flow that marked the end of the cycle in question); the remaining days (the follicular phase) were divided into two equivalent time periods (Blocks 4 and 5) for each subject. Thus Blocks 1, 2, 3, 4, and 5, numbered in reverse chronological order, described comparable time periods during one menstrual cycle of each subject, regardless of its length. Blocks of data over time were analyzed using a repeated measures ANOVA (Hull & Nie, 1981); statistical significance was set at the .05 level.

Results and Discussion

Descriptive statistics for the sample are shown in Table 1. The mean cycle length (28.58 days) was slightly higher than the 28 days considered typical (Guyton, 1981), while mean % fat was within the range of values reported as typical for sedentary women (McArdle, Katch, & Katch, 1981). Data from the first day of measurement and from the same day of the next menstrual cycle (overlap measures) are included in Table 1. Results indicated that, on

Table 1. Means and standard deviations of body composition measures, age, and menstrual cycle length

Variable	Overall M (SD)	Initial M (SD)	Final M (SD)
Body weight (kg)	59.30 (6.90)	59.33 (6.71)	59.31 (6.89)
Density (g/cc)	1.047 (.0120)	1.047 (.012)	1.047 (.012)
% fat	22.97 (5.42)	22.87 (5.28)	22.94 (5.50)
Age (years)	27.15 (6.30)		
Cycle length (days)	28.58 (4.35)		
No. of body weight measures	12.7 (1.77)		
No. of hydrostatic measures	6.7 (1.12)		

Note. Initial and final measures were taken on the same days of two successive menstrual cycles; N = 26.

the average, little overall change occurred during the menstrual cycle. Test-retest correlations for these overlap measures were $r = .995$ for BW and $r = .983$ for both D and % fat, indicating very high reliabilty.

Averaged data taken during Blocks 1, 2, 3, 4, and 5 are reported in Table 2. Intercorrelations among measures of BW, D, % fat, FW, and FFW ranged between $r = .980$ to $r = .998$ across all blocks. These correlations are higher than they otherwise might have been because data was averaged within blocks for each subject, giving a better estimate of the subject's true score for that time period. However, it suggests that there is a very high correspondence between measures taken at different times in the cycle.

Repeated measures ANOVA of BW over blocks resulted in a significant F-ratio ($F = 3.65$; $p = .008$). Tukey's Honest Significant Difference Test (Kirk, 1969) indicated that the only pairwise comparison which achieved significance was Block 1 versus Block 4. The finding of a significant increase in BW during the 5 days preceding menstruation (Block 1) is consistent with findings that others have linked to water retention (Bruce & Russell, 1962; Good, 1978; Thorn et al., 1938) and/or appetite (Dalvit, 1981).

Successive runs of repeated measures ANOVA over Blocks for D, % fat, FW, and FFW were not significant. Byrd and Thomas (1983) also found no significant differences in D or % fat when 12 women were measured in each of two menstrual cycles. If the weight gain observed in the present study were due to increased hydration, then one might expect that D would decline on the basis of evidence from Girandola et al. (1977). But if the weight gain were due to increased appetite, then changes in D would be hard to predict unless caloric intake and expenditure were monitored during this time period, which

Table 2. Means and standard errors of body composition values during different time blocks of a menstrual cycle[a]

Variable	5	4	3	2	1
Body weight (kg)					
M	59.16	59.02*	59.31	59.26	59.50*
(SE)	(1.35)	(1.39)	(1.40)	(1.38)	(1.41)
Density (g/cc)					
M	1.0471	1.0469	1.0464	1.0465	1.0469
(SE)	(.0024)	(.0024)	(.0024)	(.0024)	(.0024)
% fat					
M	22.82	22.92	23.11	23.07	22.93
(SE)	(1.06)	(1.08)	(1.10)	(1.09)	(1.11)
Fat weight (kg)					
M	13.68	13.71	13.89	13.84	13.83
(SE)	(.84)	(.86)	(.88)	(.85)	(.89)
Fat free weight (kg)					
M	45.53	45.30	45.38	45.42	45.66
(SE)	(.91)	(.95)	(.92)	(.96)	(.95)

[a]Blocks 1 to 3 represent three successive, 5-day periods counting backward from menstruation; Blocks 4 and 5 represent equivalent time periods formed from the remaining data for each of 26 subjects; *$p < .01$.

they were not. It is possible that changes in hydration and appetite may have had offsetting effects, one reducing D and the other increasing D at the same time. Another possibility is that fluctuations in diet, which were not controlled in the present study, may have contributed to greater within-subject variability in all time blocks, thereby masking any underlying changes which may have been occurring due to menstrual function. Bruce and Russell (1962) found such a compounding influence of diet on menstrually related weight changes.

It should be noted that subjects were measured for BW on 44.3% of the days and for HW on 23.5% of the days in their cycle, on the average. Further, the design of the study insured that HW was performed at the times of highest and lowest observed weights. At such extremes the probability is greater for finding differences in body composition than when BW is closer to the subject's average BW. The conditions under which the subjects were observed throughout this study were realistic in a clinical sense. Therefore, it is concluded that, although consistent weight changes do occur in healthy women during the menstrual cycle, measures of D and % fat assessed via HW do not vary significantly.

The above conclusion is drawn only from group averages. Individual differences exist and in some cases are large. Within subjects, peak changes in BW averaged 1.55 kg (SE = .09). Corresponding figures for D and % fat were D = .0037 g/cc (SE = .0004) and % fat = 1.66% (SE = .17). One subject had a peak difference in % fat estimates of 3.95% fat units at two different points in her cycle. For another subject, the peak difference was only .39% fat units. Individual differences in this area appear to be a topic which warrants further research.

References

Brozek, J., Keyes, A., & Anderson, J. (1963). Densitometric analysis of body composition: Revision of some quantificative assumptions. *Annals of the New York Academy of Science, 110*, 112-140.

Bruce, J., & Russell, G.F. (1962). Premenstrual tension: A study of weight changes and balances of water, sodium and potassium. *The Lancet, 2*, 267-271.

Byrd, P.J., & Thomas, T.R. (1983). Hydrostatic weighing during different stages of the menstrual cycle. *Research Quarterly, 54*(3), 296-298.

Dalvit, S.P. (1981). The effect of the menstrual cycle on patterns of food intake. *American Journal of Clinical Nutrition, 34*, 1811-1815.

Girandola, R.M., Wiswell, R.A., & Romero, G. (1977). Body composition changes resulting from fluid ingestion and dehydration. *Research Quarterly, 48*(2), 299-303.

Good, W. (1978). Water relations of the ovarian cycle. *British Journal of Obstetrics and Gynecology, 85*, 63-69.

Guyton, A.C. (1981). *Textbook of medical physiology* (6th ed.). New York: W.B. Saunders.

Hull, C.H., & Nie, N.H. (1981). *SPSS update 7-9*. New York: McGraw Hill.

Kirk, R.E. (1969). *Experimental design: Procedures for the behavioral sciences*. Belmont, CA: Brooks/Cole.

Lohman, T.G. (1981). Skinfolds and body density and their relation to body fatness. *Human Biology, 53*(2), 181-225.

McArdle, W.D., Katch, F.I., & Katch, V.L. (1981). *Exercise physiology*. Philadelphia: Lea & Febiger.

Novak, L.P. (1974). Analysis of body compartments. In L.A. Larsen (Ed.), *Fitness, health, and work capacity: International standards for assessment* (pp. 241-255). New York: MacMillan.

Siri, W.E. (1961). Body composition from fluid spaces and density: Analysis of methods. In J. Brozek & A. Henschel (Eds.), *Techniques for measuring body composition* (pp. 223-244). Washington, DC: National Academy of Science.

Thorn, G.W., Nelson, K.R., & Thorn, D.W. (1938). A study of the mechanism of edema associated with menstruation. *Endocrinology, 22*(2), 155-163.

Wilmore, J.H., Vodak, P.A., Parr, R.B., Girandola, R.N., & Billing, J.E. (1980). Further simplification of a method for determination of residual lung volume. *Medicine and Science in Sports and Exercise, 12*(3), 216-218.

24

Hydrostatic Weighing Without Head Submersion

Joseph E. Donnelly and Stephanie Smith Sintek
KEARNEY STATE COLLEGE
KEARNEY, NEBRASKA, USA

Hydrostatic weighing (HW) has generally been considered the most accurate noninvasive method of determining body composition. Investigators have studied various body positions and breathing maneuvers to allow for a more accurate procedure that is comfortable for subjects. Regardless of body position or breathing maneuver, standard HW requires that the subject be completely submerged and remain submerged for a period of time. It becomes apparent that some subjects do not like to submerge their heads and some even dislike facial contact with water. This apprehension regarding total submersion makes results for these subjects suspect and may completely prevent use of HW.

Residual lung volume (RV) is most often utilized to account for buoyancy forces of trapped air. It is generally believed that the smallest lung volume will be the least affected by hydrostatic pressure and is therefore the most appropriate volume to utilize (Welch & Crisp, 1958). Although RV may be the most appropriate volume to utilize in HW, other investigators have shown that larger lung volumes sacrifice little accuracy and are more comfortable for the subject. Weltman and Katch (1981) compared HW at total lung capacity (TLC) as an alternative method to HW at RV. They reported negligible differences in the percentage of body fat obtained between the two methods and noted that the subjects appeared more comfortable. In 1980, Thomas and Etheridge compared HW at RV and functional residual capacity (FRC) and reported no significant differences in body composition between the methods. Additionally, it was noted that the subjects were more comfortable which produced a more stable and more easily read scale weight.

In an effort to ensure the comfort of the subject and therefore obtain a more reliable lung volume, Donnelly, Sintek, Anderson, and Pellegrino (1984) introduced a pilot study which described a method of HW in which the subject was not required to submerge the head. Hydrostatic weighing at total lung capacity submerged (TLC_S) was compared to HW at total lung capacity without head submersion (TLC_{NS}). A correlation of .967 was found between the estimates of percent fat produced by the two methods for the original 15 subjects. A second group of 8 subjects was hydrostatically weighed by the two methods and a correlation coefficient of .980 was found between the two sets of percent fat estimates. Subjects commented that not having to submerge the head reduced anxiety considerably. This investigation duplicated the previous pilot study of Donnelly et al. (1984) with a larger population and refined the techniques necessary to produce accurate estimations of body composition using HW without head submersion.

Method

The subjects were 40 male volunteers between the ages of 18 and 34. All subjects were questioned regarding pulmonary disease, which would eliminate participation in the study. Each subject was weighed on a calibrated Detecto Doctors' Scale to the nearest 1/4 pound. In order to use standard body density equations for the calculation of body composition, it was necessary to generate a regression equation to predict the difference in underwater weight observed between the two methods. It was suggested that these differences would be accountable largely to head volume and that an estimation of head volume could be obtained by measurement with a slide caliper. Head width was determined with a horizontal measure at the level of the superior ear. Head length was determined with an oblique measure from a point at midmandible to a point on the superior skull even with a vertical line from the anterior ear.

The subjects were hydrostatically weighed at TLC_S and TLC_{NS} using a Chatillon autopsy scale accurate to 25 g. Water temperature was maintained between 32° and 34° C for all subjects. Residual volume was determined from vital capacity (VC) as described by Wilmore (1969) while the subject was seated in water and breathing into a 13.5 l Collins respirometer. Total lung capacity was calculated as VC and RV. The subjects were administered five trials of each method. Submersion for TLC_S was according to standard procedures. The subject inhaled maximally and lowered his head so that submersion was complete. To perform the TLC_{NS} method, the subject inhaled maximally and submerged such that the water line was at the inferior surface of the chin and the inferior border of the ear lobe.

The criterion weights for TLC_S were the average of the two lowest trials. The criterion weights for the TLC_{NS} method were the average of the middle three trials with the lowest and highest trials eliminated. This procedure was thought to minimize error by improper positioning of the subject either too high or too low in the water. Body density was calculated according to the equation of Goldman and Buskirk (1961). Percentage of body fat was determined by the equation of Brozek, Grande, Anderson, and Keys (1963).

In order to use the weights obtained during TLC_{NS}, it was necessary to use a "correction factor." This was accomplished by deriving a multiple regression equation using head length and width as independent variables and differences in observed underwater weight between the methods as the dependent variable. The equation for determining the correction factor is

$$Y = 4605.94 - 247.56X_1 - 182.93X_2$$
$$Y = \text{correction factor (gm)}$$
$$X_1 = \text{head width (cm)}$$
$$X_2 = \text{head length (cm)}$$
$$SEE = 269.92$$
$$N = 40$$

As shown in the example:

Width $= 15.1$ cm
Length $= 22.6$ cm
$Y = 4605.94 - 247.56X_1 - 182.93X_2$
$Y = 4605.94 - 247.56 (15.1) - 182.93 (22.6)$ or
$Y = -3266.4$

Y would be used to correct underwater weight obtained during TLC_{NS} back to a value expected if TLC_S was performed. Normal equations for percentage of body fat may then be used.

An independent group of 11 male volunteers between ages 18 and 34 was tested using identical procedures to determine the ability of the regression to predict a correction factor in subjects outside the population from which the regression was derived. Correlation coefficients were used to compare the predictions given by the TLC_S and TLC_{NS} methods in the treatment group of 40 subjects and the independent group of 11. Repeated measures ANOVA were used to determine whether significant differences occurred between the methods in both groups.

Results

The regression of head measurements on the difference between observed underwater weights of the TLC_S and the TLC_{NS} methods enabled a correction factor to be generated when head width and length measurements were placed in the equation. This correction factor allowed for the prediction of underwater weight with the head submerged and the subsequent use of standard equations to predict body density and fat.

Table 1 shows the results of the repeated measures ANOVA and indicates that there were no differences between the two methods with either group of subjects. Correlation coefficients (Weber & Lamb, 1970) were calculated between the body fat results of the TLC_S and TLC_{NS} methods of both groups and were shown to be .924 and .984 for the treatment and independent groups, respectively.

Table 1. Comparison of 2 hydrostatic weighing techniques (TLC$_s$ and TLC$_{NS}$) for 2 groups of male volunteers

Source	SS	df	MS	F
Treatment group (N = 40)				
Subjects	1071.2	39		
Treatments	4.72E-06	1	5.72E-06	4.85E-06[a]
Error	45.92	39		
Total	1117.1	79		
Independent group (N = 11)				
Subjects	1010.1	10		
Treatments	1.02	1	1.2	1.34[a]
Error	9.1	10	.90	
Total	1020.3	21		

Note. TLC$_s$ = total lung capacity with head submerged; TLC$_{NS}$ = total lung capacity without head submerged; [a]not significant (p > .05).

Discussion

The results of this investigation indicate a high correlation between body fat measures calculated from TLC$_s$ and TLC$_{NS}$ methods. Furthermore, the ANOVA indicates no significant differences in estimates of percentage of body fat between the two methods. The results from the independent group of subjects suggest that the correction factors generated from the regression were capable of predicting with good accuracy the observed underwater weights when the subjects had submerged the head. Because the subjects exhibited a limited age range, it is questionable whether the regression is applicable to the general population. It is most probable that additional equations will have to be generated for younger subjects who are still growing.

The subjects commented that the TLC$_{NS}$ method was the more comfortable of the two methods. Having the head above water reduced anxiety because the subject had immediate access to air. This allowed the subjects to hold their breath to the last possible moment and resulted in very stable scale readings. It would be possible to compare the submerged versus the nonsubmerged methods at RV, but that would appear contradictory to the objective of making the subject more comfortable. Weltman and Katch (1981), Thomas and Etheridge (1980), and Donnelly et al. (1984) have reported that the use of larger lung volumes than residual volume is more comfortable.

The TLC$_{NS}$ procedure correlated highly with the TLC$_s$ method in both the treatment group and the independent group (r = .92, r = .98). The standard error of estimate was 1.5 and 1.4 percent body fat for treatment and independent groups, respectively, and represents a normal range of measurement error (Brozek et al., 1963). Hydrostatic weighing without head submersion is an alternative to total submersion methods for subjects who exhibit anxiety. Furthermore, the use of the nonsubmerged method may have considerable application in populations such as children, older adults, and the handicapped who may be difficult to submerge.

References

Brozek, J., Grande, F., Anderson, T., & Keys, A. (1963). Densitometric analysis of body composition: Revision of some quantitative assumptions. *Annals of the New York Academy of Science,* **110**, 113-140.

Donnelly, J.E., Sintek, S.S., Anderson, J.T., & Pellegrino, L. (1984). Hydrostatic weighing without head submersion. *Proceedings of the 94th Nebraska Academy of Sciences* (p. 20).

Goldman, R.F., & Buskirk, E.R. (1961). Body volume measurement by underwater weighing: Description of a method. In J. Brozek & A. Henschel (Eds.), *Techniques for measuring body composition* (pp. 78-89). Washington, D.C.: National Academy of Science.

Thomas, T., & Etheridge, G. (1980). Hydrostatic weighing at residual volume and functional residual capacity. *Journal of Applied Physiology: Respiratory, Environmental, and Exercise Physiology,* **49**, 157-159.

Weber, J., & Lamb, D. (1970). *Statistics and research in physical education.* St. Louis: C.V. Mosby.

Welch, B.D., & Crisp, C.E. (1958). Effect of level of expiration on body density measurement. *Journal of Applied Physiology: Respiratory, Environmental, and Exercise Physiology,* **12**, 399-402.

Weltman, A., & Katch, V. (1981). Comparison of hydrostatic weighing at residual volume and total lung capacity. *Medicine and Science in Sports and Exercise,* **13**, 210-213.

Wilmore, J.H. (1969). The use of actual predicted and constant residual volumes in the assessment of body composition by underwater weighing. *Medicine and Science in Sports,* **1**, 87-90.

25

Bilateral Symmetry and Reliability of Upper Limb Measurements

James A.P. Day
THE UNIVERSITY OF LETHBRIDGE
LETHBRIDGE, ALBERTA, CANADA

The orthodox method of taking upper limb measurements, herein called the projected lengths method, is vulnerable to error from two important sources: the anthropometrist and the subject. The error contributed by the subject was the focus of this investigation.

The four height measures (acromiale, radiale, stylion, and dactylion heights) required for calculation of the projected lengths of the upper limb segments should be taken with the subject in a stable, standardized pose, erect with weight equally on each foot, and arms and hands extended downward maximally. Any departure from the standard pose during the measurement contributes to the "error" in the observed measurement. The error can be defined as the difference between the observed measure and the "true" measure.

Because the projected length of each limb segment is calculated by subtracting the height of one landmark from the height of another, the errors in both measures contribute to the difference between the observed measure and the true measure. It is possible that the errors in both height measurements can be in the same direction and of the same magnitude, in which case the observed length and the true length would be identical. It is also possible that the two errors can be in opposite directions, resulting in an observed length which differs from the true length by the algebraic difference between the two errors.

Reliability (temporal stability) is well recognized as desirable in anthropometric measurement. The Pearson correlation coefficient, as an index of reliability, is very high and positive when most anthropometric measures are taken by an experienced investigator. When random errors occur in measurement, one might be tempted to assume that the obtained reliability coefficient could be affected randomly, having an equal chance to be higher or lower than the true value. In fact, the usual effect of random errors is to cause the obtained correlation to be closer to zero than the real correlation, and the chance that an obtained correlation would be higher (i.e., farther from zero) than a true correlation is small. This probability approaches zero as the true correlation approaches 1.000. From the foregoing, it should follow that when two measurement techniques are compared, the technique with the greater reliability (i.e., the higher test-retest correlation coefficient) is the technique of choice.

As humans we demonstrate a high degree of bilateral symmetry in our upper limb lengths. Despite this fact we would not expect to find perfect symmetry in a group of unselected subjects. It is reasonable, however, to anticipate very high positive correlations between contralateral limb segments. Thus when bilateral symmetry is assessed using two measurement techniques, the technique showing the highest "coefficients of symmetry" can be judged less vulnerable to error and should be the technique of choice.

Method

Fifty-three male and female university undergraduates were marked and measured by a single anthropometrist, according to the projected lengths procedures described by Ross and Marfell-Jones (1982). The measures were taken on both arms. The marks were then used to measure the segments with an anthropometric tape. In this procedure the zero mark of the tape was held at the acromiale, and successive readings were taken at the radiale (#1), stylion (#2), and dactylion (#3). For each subject the entire procedure was repeated 4 to 10 days later.

Four projected lengths were calculated from the projected heights, as follows:

Upper Arm Length = Acromiale Height − Radiale Height
Forearm Length = Radiale Height − Stylion Height
Hand Length = Stylion Height − Dactylion Height
Upper Extremity Length = Acromiale Height − Dactylion Height

The following four comparable measures were attained with the tape:

Upper Arm Length (tape) = Measure #1
Forearm Length (tape) = Measure #2 − Measure #1
Hand Length (tape) = Measure #3 − Measure #2
Upper Extremity Length (tape) = Measure #3

Means and standard deviations were calculated for all measures for both test administrations. Thirty-two Pearson correlations were calculated. Sixteen of these were test-retest correlations, for example, Right Upper Arm Length (first test) versus Right Upper Arm Length (second test). The other 16 were bilateral correlations, for example, Right Upper Arm Length (second test) versus Left Upper Arm Length (second test).

Results

In Table 1 are presented the means, standard deviations, minimums, and maximums for all of the length measures for the 53 subjects. In all cases except

Table 1. Means, standard deviations, and range values for upper limb length measures (cm)

Measures	Projected lengths								Tape lengths							
	Test				Retest				Test				Retest			
	M	SD	Min.	Max.	M	SD	Min.	Max.	M	SD	Min.	Max.	M	SD	Min.	Max.
Upper arm length (right)	32.84	2.16	28.8	37.4	33.15	2.05	28.9	37.1	33.76	2.11	29.5	38.4	33.95	1.98	30.3	38.0
Forearm length (right)	24.92	1.62	20.6	28.5	24.43	1.73	20.3	29.0	24.36	1.43	21.1	27.8	24.19	1.57	20.7	27.4
Hand length (right)	17.69	1.21	14.0	20.9	18.13	1.19	15.4	20.6	18.70	1.10	16.1	20.8	18.73	1.07	16.2	20.7
Upper extremity length (right)	75.44	4.22	64.5	82.9	75.70	4.22	65.3	83.0	76.82	4.15	67.0	84.5	76.87	4.05	67.3	84.7
Upper arm length (left)	32.90	2.04	28.6	37.4	32.96	2.01	28.2	37.2	33.64	2.01	29.3	38.3	33.60	2.12	28.4	38.2
Forearm length (left)	23.50	1.82	18.5	28.3	23.66	1.77	19.4	27.6	24.17	1.53	21.0	27.9	24.18	1.54	20.8	27.1
Hand length (left)	18.44	1.27	15.3	20.6	18.56	1.21	15.7	21.0	18.70	1.15	15.8	20.6	18.78	1.53	16.3	20.9
Upper extremity length (left)	74.83	4.12	64.0	83.1	75.17	4.23	66.1	82.5	76.51	4.14	66.8	83.9	76.56	4.31	66.0	84.7

Note. N = 53.

Table 2. Comparison of test-retest correlations of upper limb length measures for two measuring techniques

Measures	Projected length r	Tape r
Upper arm length (right)	.918	.936
Forearm length (right)	.816	.948
Hand length (right)	.636	.928
Upper extremity length (right)	.963	.984
Upper arm length (left)	.928	.930
Forearm length (left)	.764	.943
Hand length (left)	.665	.916
Upper extremity length (left)	.956	.985

for Forearm Length (Right), the mean segmental lengths measured with the anthropometric tape exceeded those obtained with the conventional technique. These mean differences varied from 0.22 to 1.01 cm. For the Upper Extremity Length means, in which these differences are summed, the tape means exceeded the projected means by 1.17 to 1.68 cm.

Table 2 presents the 16 test-retest correlations to allow comparison between the reliability coefficients for the projected lengths and the tape lengths. In each column the two highest coefficients are those for Upper Extremity Length and the two lowest are those for Hand Length. All coefficients in the Tape column exceed the comparable ones in the Projected column. For Upper Arm Length (Left) the difference is unimportant (.930 to .928), but all other comparisons show real differences in the correlations. These statistics indicate an important difference in temporal stability between the two techniques.

A comparison of the bilateral (right vs. left) correlations is presented in Table 3. In each column (as in Table 2) the largest correlations were for Upper Extremity Length and the smallest were for Hand Length. In every comparison between the columns the largest correlation is in the Tape column.

Table 3. Test-retest comparisons of bilateral correlations (right limb vs. left limb) for two measuring techniques

Measures	Projected length r	Tape r
Upper arm length (test)	.917	.927
Upper arm length (retest)	.876	.941
Forearm length (test)	.776	.912
Forearm length (retest)	.840	.908
Hand length (test)	.665	.903
Hand length (retest)	.685	.898
Upper extremity length (test)	.957	.972
Upper extremity length (retest)	.959	.981

Discussion

The projected lengths technique is vulnerable to "subject error" because of the subject's natural tendency to sway slightly when standing erect. The clear advantage of the tape technique used here is that when the subject moves, the zero point of the tape also moves. The clear disadvantage of the tape technique is the tendency to overestimate the Upper Extremity Length and the segmental lengths. A true length measure should be a straight line and the tape is not straight as it conforms to the upper limb's contours. In the first few centimeters below the acromiale, this problem is most pronounced as the tape bends over the deltoid muscle.

For the projected lengths technique 4 out of 8 test-retest correlations and 5 out of 8 bilateral correlations were below .900. For the tape technique described, all 16 correlation coefficients exceeded .900. The projected lengths technique should be modified to accommodate the concerns expressed here or be replaced by a more reliable technique.

References

Ross, W.D., & Marfell-Jones, M.J. (1982). Kinanthropometry. In J.D. MacDougall, H.A. Wenger, & H.J. Green (Eds.), *Physiological testing of the elite athlete* (pp. 75-115). Ottawa: Mutual Press.

26

The Use of Nuclear Magnetic Resonance Imaging (NMRI) for Biomechanical Parameter Acquisition in Functional Movement Studies

Herman J. Woltring
EINDHOVEN, THE NETHERLANDS

An important goal of quantitative motion studies in a sports training, rehabilitative, or orthopedic context is to assess the force distributions in the neuromusculoskeletal (NMSK) system. Knowledge of these forces is important in view of many pathologies in the NMSK system that are thought to be correlated with biomechanical factors. Examples are tendon and ligament lesions in sports, degeneration of the joints' articulating surfaces during aging, and loosening or migration of artificial joints some years after a total joint replacement. Bone structure is known to adapt to dominating forces exercised within the bone during typical activities of daily living (Wolff's law); knowledge of these force distributions may help in predicting both useful and inappropriate changes within the bones over time.

Unfortunately, internal forces cannot be measured directly without surgical intervention: Only in studies of total joint replacement are research results on pressure distributions at the implant-bone interface beginning to become available, following the pioneering work of Rydell (1966). In other situations, indirect methods must be used; *electromyography* and the *inverse dynamics* approach are two complementary examples. The finite element method of structural mechanics (Huiskes & Chao, 1983) is a highly useful tool for deter-

This investigation was sponsored by Philips Medical Systems, Eindhoven, The Netherlands.

mining interior load distributions, but it relies heavily on the assumed loading conditions at the interfaces with the environment.

In electromyography, the forces developed by individual muscles or muscle groups are estimated from their electrical activity as measured by surface electrodes on the skin or by indwelling electrodes (Hof, 1984). The inverse dynamics approach relies on Newton's laws; here, the net forces and moments at the joints can be evaluated from the displacements of body segments and their time derivatives, in combination with measurement of environmental interaction forces (from force plates) and of the mass distributions within the body segments (Cappozzo, 1984; Hatze, 1984; Woltring, 1984). The net forces and moments at the joints can be useful in their own right (Winter, 1984), but they provide no information on the amount of cocontraction between antagonistic muscles, or about the loading of the ligaments, tendons, and articulating surfaces. Recent research views the muscles not only as active moment generators at the joints, but also as joint stabilizers in posture and movement control (Hogan, 1984). Here, the total, absolute sum of the cocontracting forces is involved; this sum cannot be evaluated from the inverse dynamics in isolation.

A combination of electromyography and inverse dynamics is not sufficient, however. The mechanico-mathematical models currently being designed and validated require geometric and mechanical information on joint components such as attachment points and material properties of ligaments and tendons, and the geometry and lubrication properties of the articulating surfaces. These data are usually estimated from a limited database consisting of cadaveric measurements; it remains to be seen to what extent they apply to the live subject. This paper presents an argument for the use of contemporary medical imaging technology, particularly of computerized tomography (CT), for acquiring the necessary geometric information. Emphasis is given to the potential of Nuclear Magnetic Resonance Imaging (NMRI) for this purpose.

Medical Imaging

Diagnostic imaging in medicine started with what today is called conventional X-ray imaging; since the 1960s, a variety of CT approaches has revolutionized this field (IEEE, 1983). In CT, a large number of *projections* along different lines or in different planes are used for mathematically reconstructing the distribution of some physical property of the biological material being studied. In X-ray CT, this is the absorption coefficient for X-rays; in ultrasonic CT, the absorption coefficient and propagation velocity of sound; in Nuclear Magnetic Resonance (NMR), the distribution of water and, to a lesser extent, of other materials (P, Na, F) in the organism. Each of these modalities has its own advantages and shortcomings in terms of imaging potential, patient risk and exposure, acquisition and processing time, and cost.

X-ray CT, for example, is excellent for imaging the bones and various soft tissue structures, and it has a relatively high spatial resolution (less than 1 mm in joint imaging). However, only transverse scans can be practically made, and X-ray exposure should be minimal. Furthermore, measurements in the neighborhood of metallic implants are problematic because many implants exhibit too much X-ray absorption for proper functioning of conventional reconstruction algorithms; in this case, special and often time-consuming algorithms are required in order to recover the bone structure in the implant's environment (Seitz, 1984).

Ultrasound is useful for soft tissue imaging; it is cheap and can be used to obtain arbitrarily oriented "slices" of the organism. However, data are lost at the bones because ultrasound cannot pass through this material, multiple reflections of the sound waves aggravate the uniqueness of the image, and resolution is usually rather limited.

Nuclear Magnetic Resonance is the newest imaging modality; a good survey is contained in the special issue on NMR Imaging and Spectroscopy of the *British Medical Bulletin*, April 1984. NMR exploits the property of certain atoms, hydrogen in particular, to perform a precessing movement in a magnetic field when they are disturbed from a stationary state by means of an *rf* electromagnetic impulse. The precessing movement occurs at a frequency which is proportional to the local magnetic field strength, and it decays at a time rate which is specific for the chemical structure; this decay can be measured as an *rf* signal. In fact, there are two different decay times (T1 and T2) that together with the proton density ϱ can be used for imaging purposes. T1 refers to a decay component parallel to the magnetic field; T2 refers to a component perpendicular to the field direction.

NMRI is an excellent tool for imaging soft tissue structures. There are no known health hazards at the currently practiced field strengths (except for *rf* heating of metallic implants by the proton-exciting *rf* transmitter field), images can be generated in arbitrary planes, and the outline of the bones can also be imaged. Originally it was thought that NMRI would have no use in NMSK diseases because hard bone tissue cannot be imaged and magnetic disturbances caused by metallic implants would render NMRI useless for implant studies. However, the humid parts of the bones can be reliably imaged (in this way, avascular necrosis in the femoral head has been diagnosed), and some implant materials have no magnetic properties, thus allowing the use of NMRI at the implant interface (Berquist, 1984). A variety of promising results has been published by the San Francisco group on the use of NMRI in NMSK situations (Moon et al., 1983). At Philips Medical Systems, recent research in collaboration with the University of Nijmegen has resulted in images of the knee joint *in vivo* where the cross-ligaments and menisci could be clearly distinguished. Once this information is quantized, it can be used for acquiring parameter values in movement studies. It is in this regard that research on processing of image data is necessary, because the amount of work incurred with manual processing of image data is too high to be practical.

Postprocessing Aspects

Quantitative structure identification from medical images is a field of considerable interest not only for scientific purposes, but even more for automating clinical diagnostics (Connors, Harlow, & Dwyer, 1982). Human factors in image processing render the diagnostic process sufficiently inconsistent to warrant research into automating this process. The multitude of different "features" used by the human observer, and their low discernibility because of contrast, limited signal-to-noise ratio, boredom, and fatigue all affect the "human equation" in diagnostic imaging, and there is a keen interest in reducing these factors.

The recent literature on pattern recognition and computer graphics is extensively involved with image postprocessing. Important aspects are image enhancement (signal-to-noise ratio improvement, resolution/contrast enhancement), boundary detection, 3-D reconstruction from 2-D slices, 2-D and 3-D shape description, and classification. In the context of parameter acquisition for mathematical models of joint function, these developments are quite promising.

In image enhancement, the quality of the image is improved by some form of linear or nonlinear filtering. Typically, the data are assumed to be bandlimited and disturbed by wideband noise. At this processing level, no use is made of known anatomic structures, and the data are treated in a purely technical, "data-driven" way. For example, the use of Maximum Entropy techniques (e.g., Minerbo, 1979; Skilling, 1984) for image enhancement has shown some promising results.

For determining the boundaries of anatomic structures, various contour detection algorithms have been published (e.g., Seitz & Rüegsegger, 1983; Zhang & Geiser, 1984). A problem with purely data-driven contour detection is its nongradual failure when the data quality deteriorates; this is an important reason for image enhancement prior to contour detection. Particularly with low-resolution imaging modalities such as ultrasound and nuclear emission tomography, contour detection is a difficult problem. It is here that more "intelligent" approaches are being investigated, using prior knowledge of anatomical structures. Now, the purely data-driven nature of conventional contouring algorithms is complemented by the "knowledge-driven" methods of artificial intelligence. Typically, spline functions are used for this purpose. Once the surface shape of a biological structure (e.g., a bone) has been determined in this way, its geometry is known and can be used in a functional movement model. Similar automatic methods can be used for determining the attachment points of the soft and hard tissues.

At the present time, this type of image processing is largely a research topic at the university level. However, the high costs of health care encourage optimal use of the more expensive technology of the medical profession. It is to be expected that various forms of automatic image processing in medicine will become clinically possible within the next decade with benefits in related application fields such as acquiring geometric joint data in functional motion studies.

References

Berquist, T.H. (1984). Magnetic resonance imaging: Preliminary experience in orthopaedic surgery. *Magnetic Resonance Imaging, 2*, 41-52.

Cappozzo, A. (1984). Gait analysis methodology. *Human Movement Science*, **3**(1/2), 27-50.

Connors, R.W., Harlow, C.A., & Dwyer, S.J. (1982). Radiographic image analysis: Past and present. *IEEE 1982 Pattern Recognition Proceedings* (pp. 1152-1169). New York: The Institution of Electrical and Electronics Engineers.

Hatze, H. (1984). Quantitative analysis, synthesis, and optimization of human movement. *Human Movement Science*, **3**(1/2), 5-25.

Hof, A.L. (1984). EMG and muscle force: An introduction. *Human Movement Science*, **3**(1/2), 119-153.

Hogan, N. (1984). Adaptive control of mechanical impedance by coactivation of antagonist muscles. *IEEE Transactions on Automatic Control*, **AC-29**(8), 681-690.

Huiskes, R., & Chao, E.Y.S. (1983). A survey of finite element analysis in orthopaedic biomechanics: The first decade. *Journal of Biomechanics*, **16**(6), 385-409.

IEEE. (1983). Special issue on computerized tomography. *Proceedings of the IEEE*, **71**(3), 291-435.

Minerbo, G. (1979). MENT: A maximum entropy algorithm for reconstructing a source from projection data. *Computer Graphics and Image Processing*, **10**, 48-68.

Moon, K.L., Genant, H.K., Helms, C.A., Chafetz, N.I., Crooks, L.E., & Kaufman, L. (1983). Musculoskeletal applications of nuclear magnetic resonance. *Radiology*, **147**(4), 161-171.

Rydell, N.W. (1966). Forces acting on the femoral head prosthesis. *Acta Orthopaedica Scandinavica* (Suppl. 88).

Seitz, P. (1984). *Computertomographische Osteodensitometrie beim metallischen Kunstgelenk*. Dissertation ETH 7585, Federal Institute of Technology, Zürich, Switzerland.

Seitz, P., & Rüegsegger, P. (1983). Fast contour detection algorithm for high precision quantitative CT. *IEEE Transactions on Medical Imaging*, **MI-2**(3), 136-141.

Skilling, J. (1984). The maximum entropy method. *Nature*, **309**(5971), 748-749.

Winter, D.A. (1984). Kinematic and kinetic patterns in human gait: Variability and compensatory effects. *Human Movement Science*, **3**(1/2), 51-76.

Woltring, H.J. (1984). On methodology in the study of human movement. In H.T.A. Whiting (Ed.), *Human motor actions—Bernstein reassessed* (pp. 35-73). Amsterdam: North-Holland.

Zhang, L.F., & Geiser, E.A. (1984). An effective algorithm for extracting serial endocardial borders from 2-dimensional echocardiograms. *IEEE Transactions on Biomedical Engineering*, **BME-31**(6), 441-447.

27

Integrated Surface and Deep Structure Mapping of the Human Anatomy

Sheldon Baumrind
UNIVERSITY OF CALIFORNIA
SAN FRANCISCO, CALIFORNIA, USA

This paper is meant as a brief introduction to the use of photogrammetric methods in the acquisition of data on human subjects with particular attention to the problem of integrating data from different overlapping three-dimensional maps. We will consider four different but interrelated issues.

1. The general method of constructing individual three-dimensional maps from pairs of overlapping images acquired from different camera positions. (Note: the word *camera* will be used generally in this paper to describe any device used to generate and capture images. In this sense, the word *camera* includes X-ray machines, ultrasound, and infrared devices.)
2. The principle by which data from different three-dimensional maps can be merged into a single integrated three-dimensional map.
3. The problem of choosing an appropriate frame of reference.
4. The distinction between the location of discrete points and the mapping of contours.

General Method for Constructing Three-Dimensional Maps From Pairs of Images

The biological paradigm for three-dimensional mapping by integration of data from two perspectives is the binocular vision of humans and many other animals. When we view points at some distance from us, our eyes align and focus on the points. Figure 1 represents schematically the act of viewing a pair of points located at different distances from the observer. Line segments can readily be drawn from each point through the optical center of the lens of each eye. The included angle between the pair of such lines that intersects at any given spatial point is known as the parallactic angle of the point. Points at different distances from an observer have different parallactic angles. Different points lying the same distance from the observer have the same parallactic angle. Differences in parallactic angle are automatically interpreted by the brain as differences in distance.

An alternative way of measuring the parallax of a point uses linear rather than angular measurements. One can demonstrate this method by marking two small dots about an inch apart on a piece of blank paper and holding the paper about 8 in. in front of the face with the dots oriented parallel to the line connecting the pupils of both eyes. Assuming that the eyes are reasonably well balanced, one can with some strain view the two dots simultaneously, one dot with each eye. If this exercise is conducted successfully, the brain will perceive but a single point which will appear to be positioned beyond the plane of the paper (Figure 2a). If the procedure is repeated with the two dots moved slightly farther apart, the fused point will appear to be positioned farther behind the plane of the paper (Figure 2b). One can physically cut the paper in half between the two physical dots (Figure 2c). After this is done, the apparent position of the fused point can be made to displace to a lesser or greater distance beyond the plane of the paper merely by moving the two pieces of paper closer together or farther apart parallel to the line between the eyes (Figure 2d).

The geometric principle underlying this phenomenon can be used to measure distance. If the distance between the eyes and the distance from each eye to

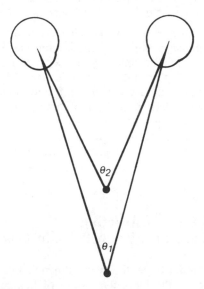

Figure 1. Schematic representation of visual parallax. Points at different distances from the viewer have different parallactic angles (θ_1 and θ_2).

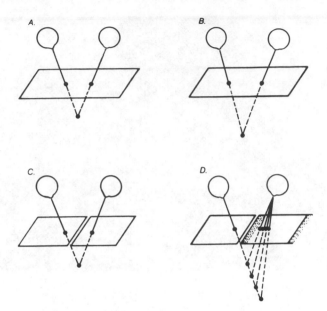

Figure 2. Measuring parallax by a linear method. (A) Two points marked on a paper appear, when viewed each with one eye, as a single virtual dot beyond the plane of the paper. (B) If the physical points on the paper are scribed further apart, the virtual image will be seen to lie farther beyond the paper. (C) The paper is cut apart to allow the distance between the points to be varied at will. (D) If the two pieces of paper are varied in their distance from each other, the virtual point will appear to lie at different distance beyond the paper. Since the distance of the virtual point beyond the plane of the paper varies systematically with the distance between the physical points on the plane, it follows that the "height" (or depth) of the virtual point can be calculated by measuring the linear distance between the pair of physical points. This operation is performed in Figure 3.

the plane of the paper are known, then the measured distance between the two dots can be used to compute the spatial position of the virtual point. As noted above, considerable eye strain results when each dot is viewed separately with only one eye. However, if an optical device known as a stereoscope is used, one can accomplish the task with relative comfort. Such a device presents one image to one eye and another image to the other eye through an optical train of lenses or mirrors (Figures 3a and b). Stereoscopes have long been used for amusement and in scientific applications to view pairs of photographs. Each matched pair of photographs (termed a stereopair) is, in effect, an infinite set of matched pairs of dots precisely like those we have just considered. When this technique is used for quantitative purposes, the distance between each matched, or conjugate, pair of image points is measured with a device known as a parallax bar (Figure 3c) that functions like the divided paper in Figure 2d.

The most widespread engineering application of this principle is in the making of three-dimensional maps using photographs taken from airplanes and satellites. In recent years, the same principle has begun to be employed widely in robotics, medicine, and sport to construct three-dimensional coordinate

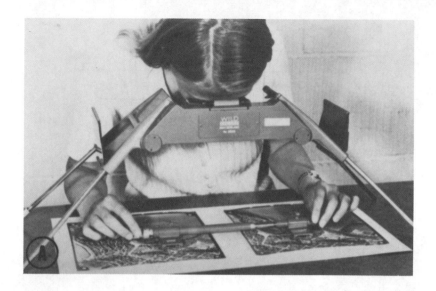

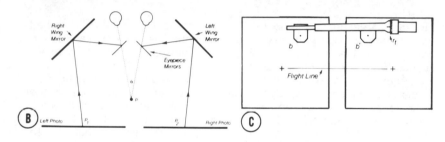

Figure 3. (A) Technician using a mirror stereoscope and parallax bar to measure the height of the point on a stereopair of aerial photographs. (B) The principle of the mirror stereoscope: When the two stereophotographs are properly oriented on the surface below the instrument, information from each photograph is transmitted on only one eye. The two eyes and the brain merge the information from the two photographs (as in Figures 1 and 2) and "perceive" a single three-dimensional virtual image in space. The heights of points within the virtual image can be measured with a parallax bar. (C) The principle of the parallax bar: This instrument consists essentially of a pair of dots scribed on separate glass plates which can be moved toward or away from each other by adjustment of a micrometer drum. The two dots function precisely like the dots on the separate pieces of paper in Figure 2, C and D. The micrometer drum is so adjusted that the two dots come to rest on the separate images of the same landmark on the two films of a stereopair. Under these conditions, the two dots on the parallax bar will appear through the stereoscope as a fused single point lying at the same height in the three-dimensonal image as does the landmark itself. The reading on the micrometer dial then represents the parallax of the point and can be used quantitatively to compute its height.

maps that can then be converted into graphic representations as desired. The basic requirements are that there be two viewing stations with a known distance between them and that each point whose three-dimensional coordinates we wish to determine is unambiguously locatable from both viewing stations.

The camera film planes may be parallel, as in the stereoscopic, or "normal" case, or they may be oriented at some angle to each other, in what is known as the convergent, or biplanar case (Figure 4). The "viewing" may be done in the visual spectrum, in the X-ray or the infrared spectrum, or even in the ultrasound or sound spectrum, although the resolution is naturally limited by the wave length of the signal employed. The target, or point being monitored, can be either active or passive. Examples of passive targets are discrete anatomical structures (such as a nose or ankle) or marks inscribed on the body of an individual being X-rayed. Active targets, that is, those that emit their own signal, are exemplified by light emitting diodes (LED), by radiobiological substances used in nuclear medicine, and by satellites and aircraft which emit signals for self-location and identification.

In general, data acquisition from active systems is much more easily accomplished and much less labor-intensive than is data acquisition from passive systems. On the other hand, the active systems require the mounting upon the subject of power supplies or other encumbrances that tend to constrain or alter performance. Some examples of specialized passive systems that our group has developed for obtaining three-dimensional information from facial and intraoral photographs and from paired X-ray films are shown in Figure 5 (Baumrind, 1975; Baumrind, Moffitt, & Curry, 1983; Curry, Moffitt, Symes, & Baumrind, 1982).

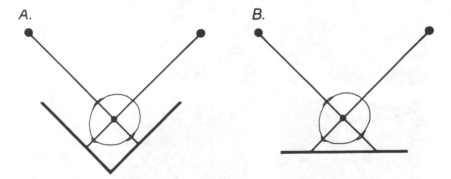

Figure 4. Diagrammatic representation of the differences in geometry between biplanar and coplanar systems for X-ray stereometry of the head. In the biplanar system (a), the film cassettes (shown as lines lying beyond the patient's head) for each of the two X-ray tubes (shown as filled dots [●]) lie in different planes. In the coplanar system (b), the two film cassettes lie in the same plane (hence, "coplanar"). In photography in the visual light range, the geometric considerations are the same except that the camera nodal point takes the place of the X-ray focal spot while the film lies on the opposite side of the nodal point from the subject instead of lying beyond the subject on the same side.

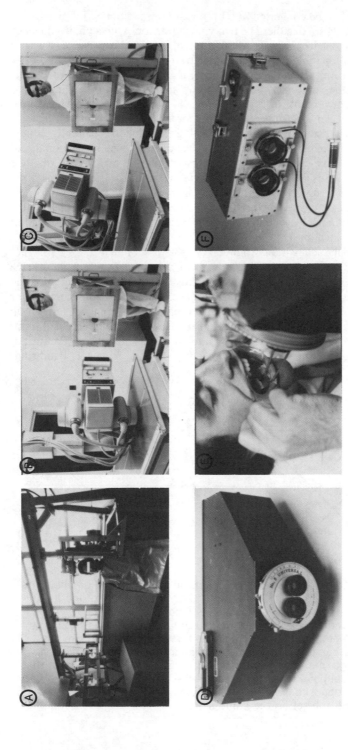

Figure 5. (A) Dedicated craniofacial X-ray system. Arrows point to twin fixed X-ray tubes. (B & C) Portable medical stereo X-ray system. The single X-ray tube is moved vertically between exposures. (D & E) Specialized stereo camera for single tooth intraoral photographs. (F) Specialized stereo camera for facial and intraoral photographs. (*Note:* The two X-ray systems are discussed in Curry, Moffitt, Symes, & Baumrind [1983]. The photographic systems are discussed in Baumrind, Moffitt, & Curry [1982].)

Merging Data From Different
Three-Dimensional Coordinate Maps

In the previous section, the task of creating *individual* three-dimensional co-ordinate maps was discussed. In many measurement situations, *multiple* maps exist and the problem is how to relate the different maps to each other. Three kinds of situations in which multiple maps are available can be identified:

1. Situations in which two or more maps are made simultaneously or nearly simultaneously from camera stations that view the subject from different perspectives but use the same narrow band of signal frequencies. Examples of this situation in the visual spectrum include conventional photographs from multiple camera stations (for measuring astronauts, for architectural mapping, and for intraoral photography in dentistry). In the X-ray spectrum, CT scanners represent a roughly equivalent technology.
2. Situations in which simultaneous or nearly simultaneous stereo informa-tion is obtained from multiple camera stations using signals of different frequencies. Examples include craniofacial mapping and analogous medical uses in plastic surgery and orthopedics in which information from photo-graphs is complexed with information from X-ray images (see Figure 6).
3. Situations in which multiple maps are not generated simultaneously but rather sequenced through time. Examples include missile tracking, automobile accident research, the study of craniofacial growth from X-rays taken at different times, gait analysis, and the monitoring of athletes in motion.

In each of these cases, individual three-dimensional maps can be made by variants of the general method outlined earlier in this paper provided that pairs of overlapping images exist that were made simultaneously from different camera positions whose locations are either known or are computable from information recorded upon the images.

In all of these cases, the problem exists of relating pairs or sets of three-dimensional maps uniquely such that the data from the several maps can be expressed in terms of a common three-dimensional frame of reference. And in all cases, the solution of the problem of relating the separate three-dimensional maps to each other is based upon the implementation of the same principle: Any pair of overlapping three-dimensional maps can be uniquely linked in a common three-dimensional frame of reference provided that both maps share three or more unambiguous points whose locations are known on each map (see Figure 7). This principle is borrowed from the field of aerial photogrammetry in which overlapping three-dimensional maps made from dif-ferent perspectives (and frequently at different scales) are routinely related to each other by means of common sets of control points (Moffitt & Mikhail, 1980).

Investigators can implement the use of this principle in any sport or medical application by placing three or more unambiguous markers on the structure being mapped. Several different implementations of this principle of tying together different maps through the use of unambiguous tie-points common to several maps are shown in Figure 8. In practice, because all such points (and indeed all other landmarks) are located with some error, it is desirable

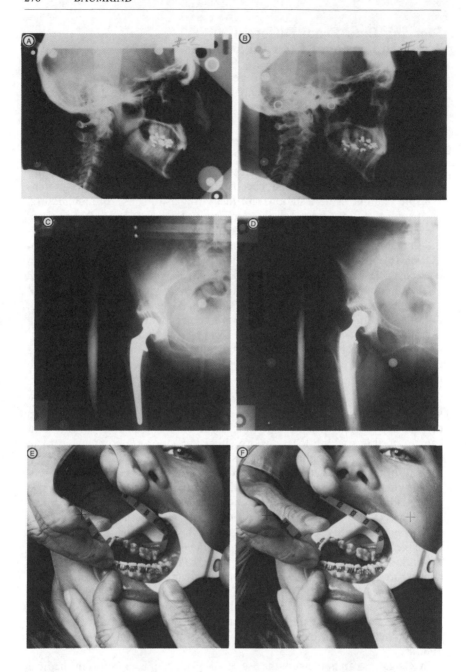

Figure 6. Representative stereo images produced by the systems of Figure 5. (a & b) Stereo skull X-ray films from the apparatus shown in Figure 5a. (c & d) Stereo hip X-ray films from the apparatus shown in Figure 5b and c. (e & f) Stereo intraoral photographs from the camera shown in Figure 5f.

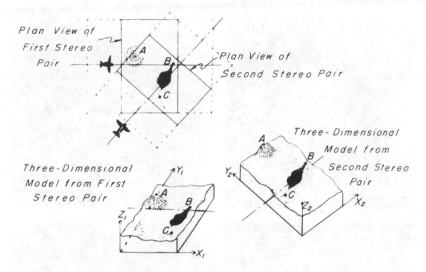

Figure 7. Graphic representation of the principle for relating data from two overlapping maps made from different perspectives. Based on the fact that points A, B, and C are known in both maps, a three-dimensional matrix rotation can uniquely position all points on both maps in terms of a common frame of reference.

that the tie-points be spaced as widely as possible in or on the structure whose performance is being monitored. It goes almost without saying that the spatial relationships among the tie-points need to be kept constant through time in order to avoid distortions in the fit between maps. It is also true that these tie-points, like the landmark location points spoken of earlier, are most easily tracked if they themselves emit detectable signals.

Choosing an Appropriate Frame of Reference

Whenever a definition of the location of an object in space is attempted, it must be done with respect to some reference system. In the construction of topographical maps (the task for which most of the methods now adapted to biostereometric use were originally devised), the choice of an appropriate reference system, variously called a frame of reference, or datum, is relatively straightforward. For such work, one almost always chooses to measure with respect to sea level. The fact that the "surface" of the earth at sea level is such an obvious datum plane for topographical mapping has exerted a very strong influence upon the instrumentation and techniques of photogrammetry such as the use of contour plotting (about which more will be said later). In the stereo analysis of biological systems, the choice of an appropriate frame of reference is usually less obvious. Unlike the terrain features which are the subject of topographical mapping, most living organisms rapidly change their

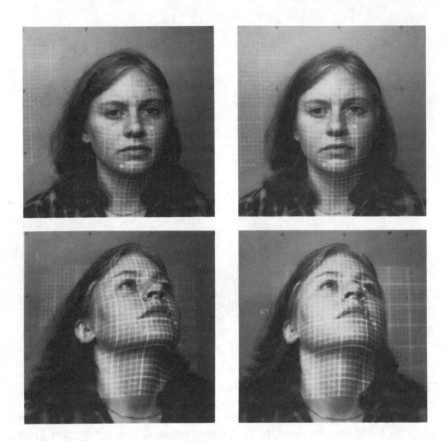

Figure 8. Four photographic stereopairs of the same individual made on the same occasion but from different perspectives. Two types of artificial discrete points have been generated on the skin surface. One is a set of lead markers on white backgrounds that maintain the same relative positions on all stereopairs and perform the same function as points A, B, and C of Figure 7 for relating these stereopairs to each other. The other type of artificial discrete point is a set of projected grid-intersections produced by special illumination procedures. These intersections are, of course, different on each map but can be merged into a single continuous map of the facial surface on the basis of the constant relationship among the lead markers. In addition, because the lead markers are also radiopaque, they can also be used to relate these facial photographs to X-ray stereo-images like those in Figure 6a and b, provided the lead markers are in place when the X-ray films are exposed.

orientations to their microenvironment. Indeed, such changes in orientation constitute the main focus of interest in kinanthropometry.

It seems to the author that there are two main datahandling tasks in the biostereometric measurement of motion. One is the merging of data obtained from different camera stations and different perspectives at the same instant in time. The other is the monitoring of the subject's motion with respect to

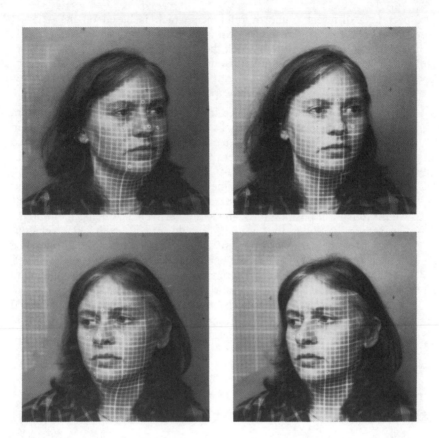

Figure 8 (cont.).

the microenvironment. Each of these tasks in fact requires a different frame of reference. For merging of data from different camera stations, a frame of reference based on the anatomy of the subject itself is usually preferred. (In the roughly analogous craniofacial growth studies with which my research group is typically occupied, we most frequently use a system in which *sella turcica* is defined as the origin, *nasion* is defined as a point on the X axis, and *anterior nasal spine* is defined as a point on the XY plane.) In the absence of X-ray information, an anatomical frame of reference could also be based on externally mounted markers such as the tie-points mentioned previously.

The use of such intraindividual frames of reference facilitates the accurate merging of different but simultaneously made three-dimensional maps, as has been discussed above. But this very process of relating all data to an internally defined frame of reference has the collateral effect of losing all information about the motion of the subject with respect to its external microenvironment (e.g., the track or gymnastic apparatus upon which an athlete is performing). In order to examine the subject's movement with respect to the external environment, one must follow through time the movement of the tie-points with respect to a different frame of reference which is based on the external environment. Such a reference frame might use as its major axes the elements of the physical apparatus upon which the subject was performing (i.e., a set of parallel bars or a high jump apparatus).

If data were gathered consistent with these principles, the following types of analytic operations would become possible:

1. At each point in time, data obtained from different perspectives and camera stations could be merged in terms of a common anatomical reference system based on the tie-points common to the several maps. The resulting integrated map would constitute a comprehensive description of the state of the subject at a single point in time.
2. Integrated maps from successive time points could be merged on their common tie-points. (This operation is based on the assumption that the tie-points would remain in position throughout the performance being monitored.) The sequence of maps thus produced would reflect the changing relationships among the various portions of the subject's anatomy (with reference to each other) during the period being analyzed.
3. Integrated maps from successive timepoints could be ordered sequentially on the basis of the changing spatial relationship between the tie-points and the external frame of reference. The sequence of maps thus produced would reflect the movement of the subject with respect to the external physical apparatus which defines the frame of reference.

The Distinction Between Discrete Point Location and the Mapping of Contours

There are two general modes in which three-dimensional data may be secured from stereopairs. One involves the location of discrete points, each of which is uniquely identified in X, Y, and Z on both images of the stereopair as explained earlier. The other mode, called the contour mode, involves the construction of a series of closed lines, each of which links together all points on the structure of interest which lie at a common "altitude" with respect to some reference plane (which is commonly known as the datum plane). In topographic mapping, contour lines are relatively easy for technicians to generate because they bypass the problem of landmark identification. On two-dimensional topographic maps that represent three-dimensional data, contour lines are particularly useful in conveying a sense of steepness. If successive contour lines fall close to each other, it is understood that the terrain elevation is changing rapidly. If, to the contrary, the contour lines are widely spaced, the observer may properly infer a situation of gradual incline. Most topographic maps combine both the discrete point and the contour methods of data representation. Uniquely identifiable landmarks, such as towns, houses, or trees (depending on the scale of the map) are indicated as discrete points. On the other hand, information on undulations and unevennesses in the ground surface is usually carried by contour lines (Figure 9).

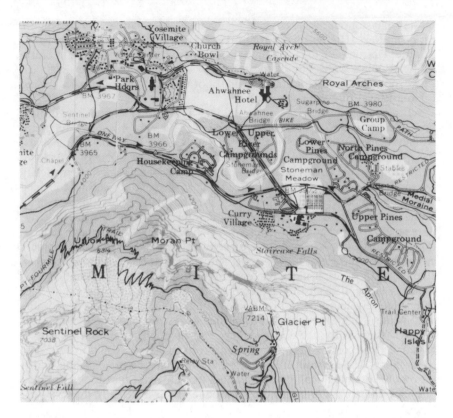

Figure 9. A portion of a combined contour and discrete point topographic map of Yosemite Valley, California.

In biostereometrics, too, the discrete point and the contour line methods of data encoding may both be employed. But two special problems obtrude which can complicate the use of contour methods for biological data. The first of these problems is that biological surfaces, either of the skin or of deeper structures such as bones and internal organs, tend to be relatively untextured and hence to be devoid of points which can be identified consistently on both images of a stereopair. This problem is particularly severe on conventional X-ray films, and as yet no practical ways are available for generating bone and internal organ contours in vivo other than the use of some form of tomography (either classical or computed).

In the case of the surface of the skin, a similar problem of lack of texture exists, but there are ways of attenuating the difficulty. Skillful lighting of the subject can accentuate small details, making it possible to produce contour lines using standard photogrammetric plotting equipment. This approach is, however, likely to be quite labor-intensive. It is therefore attractive to broadcast contour lines onto the surface being mapped at the time of photography. Moiré patterns (Karras & Tympanidis, 1982) can be projected onto the surface rather easily, or direct contour lines may be produced by sidelighting as in the "physioprint" (Sassouni, 1957) method. Alternatively, grids (Figure 6), raster lines (Hierholzer & Frobin, 1982), or random arrays of dots (Keys, Whittle, Herron, & Cuzzi, 1975) may be projected upon the surface to create artificial points which can be located by the use of discrete point methods. The problems discussed in the remainder of this section do not occur using these artificial discrete point methods.

When contour methods are used, whether with stereoplotters or by projection upon the object surface, one can encounter problems of frame of reference without being aware of the fact. This is true because with contour methods, the contour lines themselves define the orientation of a frame of reference. In topographic work, this causes no problem—the contour lines automatically lie parallel to the datum and at defined intervals above it. But in biological work, the contours simply fall in certain fixed positions relative to the camera or the lighting projector orientation with no reference to the subject anatomy. As a result, slight changes in subject orientation relative to the image acquisition geometry will produce considerable alterations in the pattern of contours cast upon the surface of the subject (Figure 10). For this reason, even when strenuous efforts are made to photograph the subject from a constant orientation, it is very difficult to make comparisons between sets of contour lines from time-sequenced maps of the same subject (and it is all but impossible to compare contour lines for different maps made from grossly different orientations). Technically, the data-handling problem is no different from what it would be if the contour data were treated as sets of discrete points to be related to tie-points as described in earlier sections. What causes confusion is that the ordered orientation of the contour lines tends to make the observer believe that they contain more biological information than is actually the case.

Conclusion

The intention of this paper has been to introduce some general problems in three-dimensional mapping that are of importance to the quantitative study of athletes in motion. Beyond these general considerations, the investigator should expect to encounter additional specific problems in the design of almost any particular biostereometric study. These specific problems will be associated with the unique geometry of the particular study and with its specific research goals. It is very important that the special measurement problems encountered in any particular biostereometric investigation be faced and mastered *before* image and data acquisition begin. In the past, many large biostereometric projects have been irreversibly compromised by poor advance planning and by erroneous technological assumptions made early on. For this reason, new investigators are advised to seek counsel from experienced biometricians and photogrammetrists prior to the initiation of image and data acquisition procedures, rather than waiting until the data reduction and data analysis stages of the study. If proper advance preparations are made, contemporary biostereometrics can make important contributions to the solution of a large number of problems in sports physiology and medicine.

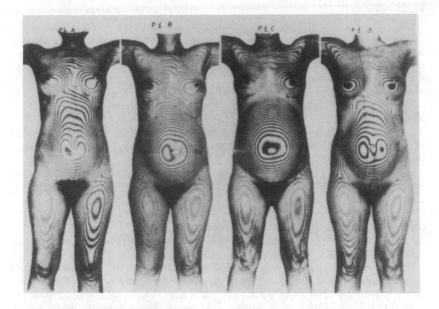

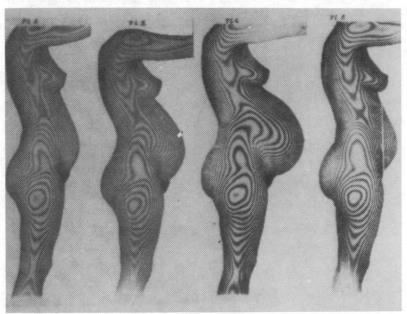

Figure 10. A set of lateral and frontal moiré pattern photographs of the same individual generated at different points in time. The positions of the contours are very sensitive to small differences in subject orientation between images. Also, without tie points such as the lead markers in Figure 8, it is very difficult to merge information from the lateral and frontal views even at the same timepoint.

References

Baumrind, S. (1975). A system for craniofacial mapping through the integration of the data from stereo x-ray films and stereo photographs. In *Proceedings of the Symposium on Close Range Photogrammetric Systems* (pp. 142-166). Urbana, IL: American Society of Photogrammetry.

Baumrind, S., Moffitt, F.H., & Curry, S. (1983). Three-dimensional x-ray stereometry from paired coplanar images: A progress report. *American Journal of Orthodontics*, **84**, 291.

Curry, S., Moffitt, F.H., Symes, D., & Baumrind, S. (1982). Family of calibrated stereometric cameras for intra-oral use. *Proceedings of the Society for Optical Engineering (SPIE)*, **361**, 7-14.

Hierholzer, E., & Frobin, W. (1982). Automatic measurement of body surfaces using rasterstereography. *Proceedings of the Society for Optical Engineering (SPIE)*, **361**, 125-131.

Karras, G., & Tympanidis, K.N. (1982). Studying abdomen size and shape variations during pregnancy: An application of moiré topography. *Proceedings of the Society for Optical Engineering (SPIE)*, **361**, 89-91. (See also other papers presented at this session, pp. 81-110.)

Keys, C.W., Whittle, M.W., Herron, R.E., & Cuzzi, J.R. (1975). Biostereometrics in aerospace medicine. In *Proceedings of the Symposium on Close Range Photogrammetry Systems* (pp. 209-220). Urbana, IL: American Society of Photogrammetry.

Moffitt, F.H., & Mikhail, E.M. (1980). *Textbook of photogrammetry* (3rd Ed.). New York: Harper & Row.

Sassouni, V. (1957). Palatoprint, physioprint and roentgenographic cephalometry as new methods in human identification. *Journal of Forensic Science, **2**, 429.

DATE DUE